Advanced Strategies for Robot Manipulators

Advanced Strategies for Robot Manipulators

Editor

Nicolas Radley

Advanced Strategies for Robot Manipulators

Edited by **Nicolas Radley**

ISBN: 978-1-68117-237-8
Library of Congress Control Number: 2016934777

© 2017 by
SCITUS Academics LLC,
www.scitusacademics.com
Box No. 4766, 616 Corporate Way,
Suite 2, Valley Cottage,
NY 10989

Notice

Reasonable efforts have been made to publish reliable data and views articulated in the chapters are those of the individual contributors, and not necessarily those of the editors or publishers. Editors or publishers are not responsible for the accuracy of the information in the published chapters or consequences of their use. The publisher believes no responsibility for any damage or grievance to the persons or property arising out of the use of any materials, instructions, methods or thoughts in the book. The editors and the publisher have attempted to trace the copyright holders of all material reproduced in this publication and apologize to copyright holders if permission has not been obtained. If any copyright holder has not been acknowledged, please write to us so we may rectify.

Preface

An industrial robot is comprised of a robot manipulator, power supply, and controllers. A robot manipulator is constructed using rigid links connected by joints with one fixed end and one free end to perform a given task (e.g., to move a box from one location to the next). The joints to this robotic manipulator are the movable components, which enables relative motion between the adjoining links. There are also two linear joints to this robotic manipulator that ensure non-rotational motion between the links, and three rotary type joints that ensure relative rotational motion between the adjacent links. Industry-specific robots perform several tasks such as picking and placing objects, movement adapted from observing how similar manual tasks are handled by a fully-functioning human arm. Such robotic arms are also known as robotic manipulators. These manipulators were originally used for applications with respect to bio-hazardous or radioactive materials or for use in inaccessible places. In industrial ergonomics a manipulator is a lift assist device used to help workers lift, maneuver and place articles in process that are too heavy, too hot, too large or otherwise too difficult for a single worker to manually handle. As opposed to simply vertical lift assists (cranes, hoists, etc.) manipulators have the ability to reach in to tight spaces and remove work pieces. A good example would be removing large stamped parts from a press and placing them in a rack or similar dunnage. Additionally, manipulator tooling gives the lift assist the ability to pitch, roll, or spin the part for appropriate placement. An example would be removing a part from a press in the horizontal and then pitching it up for vertical placement in a rack or rolling a part over for exposing the back of the part. This book, Advanced Strategies for Robot Manipulators, emphasizes on the advances in the robotic manipulation and takes a closer look at samples of robot manipulators.

Table of Contents

CHAPTER 1

Computationally Efficient Adaptive Type-2 Fuzzy Control of Flexible-Joint Manipulators

Hicham Chaoui [1], Wail Gueaieb [1,*], Mohammad Biglarbegian [2] and Mustapha C. E. Yagoub [1]

[1] School of Electrical Engineering and Computer Science, University of Ottawa, 800 King Edward Avenue, Ottawa, ON K1N 6N5, Canada

[2] School of Engineering, University of Guelph, Guelph, ON N1G 2W1, Canada

ABSTRACT

In this paper, we introduce an adaptive type-2 fuzzy logic controller (FLC) for flexible-joint manipulators with structured and unstructured dynamical uncertainties. Simplified interval fuzzy sets are used for real-time efficiency, and internal stability is enhanced by adopting a trade-off strategy between the manipulator's and the actuators' velocities. Furthermore, the control scheme is independent of the computationally expensive noisy torque and acceleration signals. The controller is validated through a set of numerical simulations and by comparing it against its type-1 counterpart. The ability of the adaptive type-2 FLC in coping with large magnitudes of uncertainties yields an improved performance. The stability of the proposed control system is guaranteed using Lyapunov stability theory.

KEYWORDS

Type-2 fuzzy control; uncertain systems; robot manipulators; flexible structures; adaptive control

1. INTRODUCTION

Flexible-joint manipulators offer several advantages with respect to their rigid counterpart, such as light weight, lower cost, smaller actuators, larger work volume, better manoeuvrability and transportability, higher operational speed, power efficiency, and larger number of applications. Thus, they are often required to operate at high speed to yield high productivity. The conflicting requirements between high speed and high accuracy make the robotic control task a challenging research problem. Reducing the weight of the arms and/or increasing the operation speed make many industrial flexible-joint manipulators face arm vibration problems, particularly in high speed motion, because of the low stiffness. This can be resolved by increasing the stiffness. However, it increases the mass, depleting the advantages listed above. Typical challenges include severe friction nonlinearities, coupling stemming from the manipulator's flexibility, varying operating conditions, structured and unstructured dynamical uncertainties, and external disturbances.

Flexibility and nonlinear friction may have some destabilizing effects when failing to compensate for modeling uncertainties in controlling flexible structures. These phenomena have been thoroughly studied in many control systems for high quality servomechanisms. Several studies show these negative consequences, such as severe tracking errors, limit cycles, chattering, and excessive noise [1,2]. Many control laws have been proposed for flexible joints [1,3,4], including classical, robust and adaptive control laws, using techniques such as singular perturbations and energy methods [5], but they generally consider (structured) parametric uncertainties only. Several models and compensation schemes have been proposed. Adaptive control techniques have been regarded among the most promising solutions to such type of problems. However, not much has been achieved yet for systems that exhibit both flexibility and severe nonlinearities.

Flexible-joint manipulators are governed by complex dynamics and hence controlling them depends on their dynamic models. There are many modeling techniques for mechanical systems, such as Lagrangian approach, Hamilton's principle, and Kane method. Yet, the system is inevitably subjected to the ubiquitous presence of high, particularly unstructured, modeling nonlinearities, such as Coulomb friction and external disturbances. The presence of such uncertainties on a manipulator driven through a flexible joint significantly changes the system's dynamics as opposed to when the load is driven with a rigid joint [6,7]. In this case, solving the inverse dynamics of the system is not realizable since the motor position is not uniquely defined at standstill. This last condition also illustrates that the actuator's state cannot be observed continuously from the load output. Henceforth, only an approximate inverse model can be realized. Therefore, modeling the system's dynamics based on presumably accurate mathematical models cannot be applied efficiently in this case. This raises the urgency to consider alternative approaches for the control of this type of manipulator systems to keep up with their increasingly demanding design requirements.

Various control techniques were proposed over the years to control flexible-joint manipulators [8,9,10]. De Luca et al. [11] and Khorasani [12] proposed feedback linearization-based controllers. However, these controllers depend on excessively noisy joint acceleration and jerk signal measurements and are hence unreliable in most real-world robotic systems. On the other hand, C. de Wit [13] proposed a robust control scheme for friction overcompensation due to uncertainties in friction models. Even as such, the suggested controllers require the full a priori knowledge of the system's dynamics. This problem has been partially overcome by several adaptive control schemes [14,15,16]. Most of these control techniques capitalize on the singular perturbation theory to extend the adaptive control theory developed for rigid bodies to flexible ones [17,18,19,20]. M. Spong [21] reduced the flexible-joint manipulators model to the standard rigid manipulators model as the joint stiffness tends to infinity. This model has been widely used by many researchers to achieve better tracking performance. For example, F. Ghorbel et al. [5] used a rigid manipulator's conventional method as slow controller and a fast feedback control law to damp out the oscillations of the joint flexibility modes. In a similar way, K. Khorasani et al. [22] illustrated how standard adaptive control schemes for rigid robots may be generalized for flexible-joint manipulators under a certain set of assumptions. Although many of these controllers are shown to be quite performant in theory, they failed to address important issues that might stand against their practical implementation, like basing the control laws on joint torques and their derivative [23,24], for instance, which are well known to be extremely noisy in real-life applications.

Moreover, such type of control algorithms uses online continuous estimation through well-defined adaptation laws of a set of the plant's physical parameters to approximate the system's dynamics. For it to provide a satisfactory performance, a typical adaptive control algorithm assumes that the dynamic model is perfectly known and free of significant external (unmodeled) disturbances. In other words, the controller is only robust to parametric or structured (also called modeled) uncertainties and possibly to minor unstructured uncertainties. Moreover, the unknown physical parameters must have constant or slowly varying nominal values. An explicit linear parameterization of the uncertain dynamics parameters also has to exist, and even if it does, it might not be trivial to derive, especially with complex dynamic systems. Although the latter condition is guaranteed for robotic systems, it might not be the case for many other dynamic models. Although some conventional adaptive control techniques, proposed in the literature, did indeed tackle external disturbance attenuation in addition to the compensation for parametric uncertainties, they did not take into consideration the effects of modeling uncertainties [14].

There is an increasing interest in developing new adaptive control schemes for robot manipulators [25,26]. For example, in [25] an adaptive neural network based sliding mode control was developed. The authors in [27] developed a neural network and an estimator to estimate the external perturbation on flexible robot manipulators. This work does not consider uncertainty in the robot parameters and was validated in simulations only. In another work, an adaptive based output-feedback control strategy for global position stabilization was developed in [28]. Although this work guarantees the adaptive regulation objective without the need for velocity feedback, it does not consider parameter uncertainties. A voltage-based adaptive control methodology for flexible joint manipulators was developed in [29,30]. The advantage of this work is that it is not torque-based control. However, it does not consider dynamic parameter uncertainties and was validated only in simulations.

In the literature, several papers on the robust control of robot manipulators have been developed, most of which use sliding mode techniques [31,32,33]. In [31] a multiple model/control strategy was proposed. It uses sliding mode control to reduce the high gain control for robot manipulators with large parameter uncertainty. In another work, a digital sliding mode controller for manipulators with three joints was developed by Corradini et al. [32]. This controller is not adaptive and uses sliding mode structure, which is prone to chattering.

On the other hand, computational intelligence tools, such as artificial neural networks and fuzzy logic controllers, have been credited in various applications as powerful tools capable of providing robust controllers for mathematically ill-defined systems that may be subjected to structured and unstructured uncertainties [34,35]. The universal approximation theorem has been the main driving force behind the increasing popularity of such methods as it shows that they are theoretically capable of uniformly approximating any continuous real function to any degree of accuracy. This has led to the recent advances in the area of intelligent control [36,37]. Various neural network and fuzzy logic models have been applied in the control of flexible joint manipulators, which have led to a satisfactory performance [9,10]. H. Chaoui et al. [38,39,40,41] used a neural network based adaptive control approach inspired by sliding mode control to learn the system's dynamics. A time-delay neurofuzzy network was suggested in [42], where a linear observer was used to estimate the joint velocity signals and eliminated the need to measure them explicitly. Subudhi et al. [43] presented a hybrid architecture composed of a neural network to control the slow dynamic subsystem and an $H\infty$ to control the fast subsystem. A feedback linearization technique using a Takagi–Sugeno neurofuzzy engine was adopted in [44]. Despite the success witnessed by neural network-based control systems, they remain incapable of incorporating any human-like expertise already acquired about the dynamics of the system in hand, which is considered one of the main weaknesses of such soft computing methodologies. In another work, an adaptive neural network based sliding mode control was developed in citeSun-2011. However, only simulations were used to verify the results. Type-1 FLCs have also been developed for robot manipulators. Some of these control techniques have hybridized type-1 FLCs with sliding mode control to achieve good robustness; e.g., Li and Huang [45] developed an MIMO adaptive sliding mode based manipulator. However, only simulations were performed to validate the theory.

As type-2 FLCs can handle uncertainties more effectively than type-1 FLCs, they are becoming a more viable tool for the control of uncertain systems [46]. Melin and Castillo [47,48] reviewed genetic algorithms, particle swarm and ant colony optimization methods in the design of optimal type-2 fuzzy systems for different applications. The authors concluded that although genetic algorithms have been used more frequently, the other two methods are rapidly gaining ground for the design of optimum type-2 FLCs. The same authors adopted the chemical reaction algorithm (CRA) to tune the parameters of type-1 and type-2 FLCs for a unicycle tracking applications [49].

While several type-1 FLCs have been developed for robot manipulators [50], very few interval type-2 FLCs have been devised for this purpose [51,52,53]. In [52], a type-2 fuzzy controller was developed and tested on a parallel robot. Most recently, in [53], a type-2 TSK fuzzy controller was used to control a modular robot. Chen developed a sliding mode based type-2 FLC for trajectory tracking. The computational complexity of the hybrid controller is rather significant because of its learning structure. In all of these works, it was concluded that type-2 FLCs can outperform their type-1 counterparts despite noise and uncertainty.

The present work contributes to the merits and the latest developments of type-2 fuzzy logic theory for the design and implementation of an adaptive type-2 FLC for the control of flexible-joint robot manipulators with uncertain dynamics. The combination of high, particularly unstructured, nonlinearities, such as in the form of Coulomb friction, and manipulator's joint elasticity changes significantly the system's dynamics. In this case, the system's inverse dynamic model cannot be found and only an approximation can be made. Therefore, modeling the system's dynamics based on presumably accurate mathematical models cannot be applied efficiently in this case. Conventional type-1 fuzzy logic systems (FLSs) can be used to identify the behavior of this highly nonlinear system with various types of uncertainties. However, type-1 fuzzy sets may not fully capture the uncertainties in the system due to membership functions and knowledge base imprecision. Hence, higher types of fuzzy sets have to be considered. However, the computational complexity of operations on fuzzy sets increases with the increasing type of the fuzzy set. Therefore, interval type-2 fuzzy sets are adopted in this work for their simplicity and efficiency to capture the severe nonlinearities of flexible-joint robot manipulators. Thus far, type-2 FLSs have been used in very few control applications, such as nonlinear control and mobile robot navigation [54]. The work presented in this manuscript has two main contributions: (i) it devises a novel adaptive control law for type-2 FLCs; and (ii) it proposes a computationally efficient inference mechanism for such type of controllers, which are reputed for their typically high computational complexity, to make them more suitable for real-time applications. This work represents one of the scarce attempts in developing adaptive type-2 FLC to control flexible-joint manipulators with uncertain dynamics. To the best of our knowledge, adaptive type-2 FLCs with a Mamdani structure have not yet been developed for the control of robot manipulators. As shall be detailed later, and unlike other types of control systems proposed in the literature, the proposed adaptive type-2 FLC is proven to be stable by Lyapunov stability theory and does not depend on the excessively noisy acceleration signals or joint torque measures [23,24]. We also present a comparative study between the proposed adaptive type-2 FLC and its type-1 counterpart to better assess their respective performances in various operating conditions.

The rest of the paper is organized as follows: Section 2 outlines the dynamical model of a typical flexible-joint manipulator. In Section 3, an overview of type-2 fuzzy logic systems is presented. In Section 4, we present interval type-2 FLSs and describe the functionality of a type-2 fuzzy inference engine. The design of the proposed controller is detailed in Section 5. In Section 6, simulation results are reported and discussed before concluding the paper with a few remarks about this important, yet complex, control problem in Section 7.

2. FLEXIBLE-JOINT MANIPULATOR DYNAMICS

2.1. Modeling of a Flexible-Joint Manipulator

The schematic representation for the ith flexible-joint in a multi-joint manipulator is shown in Figure 1. The actuator is coupled to a flexible transmission through an r:1 reduction gear. The transmission is dynamically simplified as a linear torsional spring linked directly to the load (e.g., manipulator link.)

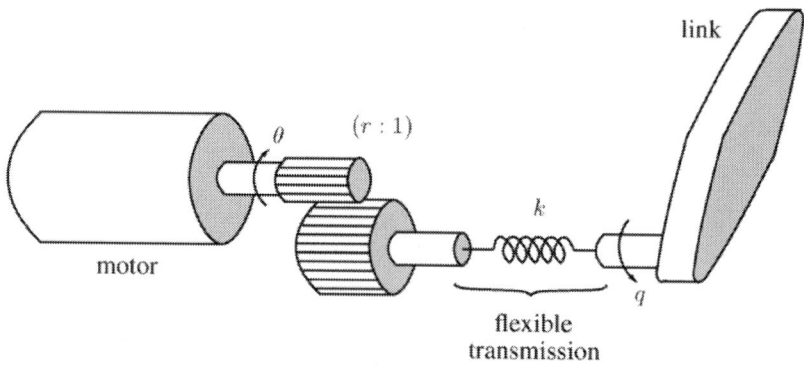

Figure 1. Flexible-joint model.

Consider a robot manipulator with n revolute flexible joints. Using Euler–Lagrange formulation, the dynamic equations of the manipulator can be written as:

$$M(q)\ddot{q} + C(q,\dot{q})\dot{q} + G(q) = \tau_t - \tau_{fl} - \tau_{dl} \tag{1a}$$

$$J_m\ddot{\theta} = \tau_m - \frac{1}{r}\tau_t - \tau_{fm} - \tau_{dm} \tag{1b}$$

$$\tau_t = K\left(\frac{\theta}{r} - q\right) \tag{1c}$$

where,

$q \in \mathbb{R}^n$:   vector of links' positions

$\theta \in \mathbb{R}^n$:   vector of motors' positions

$M(q) \in \mathbb{R}^{n \times n}$:   manipulator's positive definite inertial matrix

$C(q,\dot{q}) \in \mathbb{R}^{n \times n}$:   matrix of Coriolis and centrifugal terms

$G(q) \in \mathbb{R}^n$:   vector of gravitational torques

$\tau_t \in \mathbb{R}^n$:   vector of transmission torques

$\tau_m \in \mathbb{R}^n$:   motors' generalized torque vector (control input)

$\tau_m \in \mathbb{R}^n$:   load friction vector

$\tau_{fm} \in \mathbb{R}^n$:   motors' friction vector

$\tau_{dl} \in \mathbb{R}^n$:   load's unmodeled dynamics and external disturbance vector

$\tau_{dm} \in \mathbb{R}^n$: motors' unmodeled dynamics and external disturbance vector

$K \in \mathbb{R}^{n \times n}$: diagonal matrix of joints' stiffness coefficients

$r \in \mathbb{R}$: gear ratio

The dynamics of a robotic manipulator is characterized by the following properties:

Property 1 The inertia matrix $M(q)$ is characterized by the following properties.

(1) Positive Definite Symmetric (PDS), i.e., $MT(q)=M(q)$ and $x^T M(q)x>0$ for any non-null vector x.

(2) Upper and lower bounded, i.e., there exist two scalars $\alpha1(q)$ and $\alpha2(q)$ such that $\alpha1(q)I \leq M(q) \leq \alpha2(q)I$, where I is the identity matrix.

Property 2 The Coriolis and centripetal term $C(q,q\dot{})$ has the following properties.

(1) Matrix $\dot{M}(q) - 2C(q,\dot{q})$ is skew symmetric, i.e.,

$$x^T(\dot{M}(q) - 2C(q,\dot{q}))\,x = 0 \quad \forall\, x \in \mathbb{R}^n$$

(2) $C(q,\dot{q})\dot{q}$ is quadratic in $q\dot{}$ and bounded, i.e., there exists a scalar $\alpha3(q)$ such that $\|C(q,q\dot{})q\dot{}\| \leq \alpha_3(q)\|q\dot{}\|^2$.

Property 3 The gravity vector $G(q)$ is bounded, i.e $\|G(q)\| \leq \alpha4(q)$, for a scalar $\alpha4(q)$.

Before we proceed further, we introduce the following realistic assumption.

Assumption 1 The norm of the unknown disturbance τd is upper bounded by a scalar bd, i.e., $\|\tau d\| \leq b_d$.

2.2. Friction Modeling

Friction is highly nonlinear and it is therefore important to capture the essence of the friction phenomena with models of reasonable complexity. The behavior of friction has been extensively examined lately, a good accurate representation was introduced in [4] as the sum of Coulomb, viscous, and static friction terms. The model of such a memoryless linear or angular friction $F \in R_n$ operating along a linear or angular displacement rate vector $x^{.} = (x^{.}1, \ldots, x^{.}n)^T \in R_n$ can be expressed as

$$F(\dot{x}) = F_c \operatorname{sign}(\dot{x}) + F_v \dot{x} + F_s \Upsilon(\dot{x}) \operatorname{sign}(\dot{x})$$

$$(2)$$

where $F_c = \operatorname{diag}(F_{c_1}, \ldots, F_{c_n})$, $F_v = \operatorname{diag}(F_{v_1}, \ldots, F_{v_n})$ and $F_s = \operatorname{diag}(F_{s_1}, \ldots, F_{s_n})$ are the Coulomb, viscous and static friction positive definite diagonal matrices, respectively, and $\Upsilon(\dot{x}) = \operatorname{diag}(e^{-(\dot{x}_1/\eta_{s_1})^2}, \ldots, e^{-(\dot{x}_n/\eta_{s_n})^2})$ is a positive definite diagonal matrix representing the rate of decay of the static friction for some decay rate scalars $\eta_{s_1}, \ldots, \eta_{s_n}$, along the n degrees of freedom. The term $\operatorname{sign}(\dot{x}) \in \mathbb{R}^n$ is the vector $(\operatorname{sign}(\dot{x}_1), \ldots, \operatorname{sign}(\dot{x}_n))^T$ defined by

$$\operatorname{sign}(\dot{x}_i) = \begin{cases} 1 & , \text{if } \dot{x}_i \geq 0 \\ -1 & , \text{if } \dot{x}_i < 0 \end{cases}, \quad i = 1, \ldots, n$$

Note the friction's high sensitivity and nonlinearity in the vicinity of $x^{.} = 0$, as illustrated in Figure 2. At very low velocity, manipulators are likely to exhibit undesirable stick-slip, a cycle of stop and motion. Friction will act as a constraint and not as a motion generator [55]. τ is the control input and ε is a very small constant.

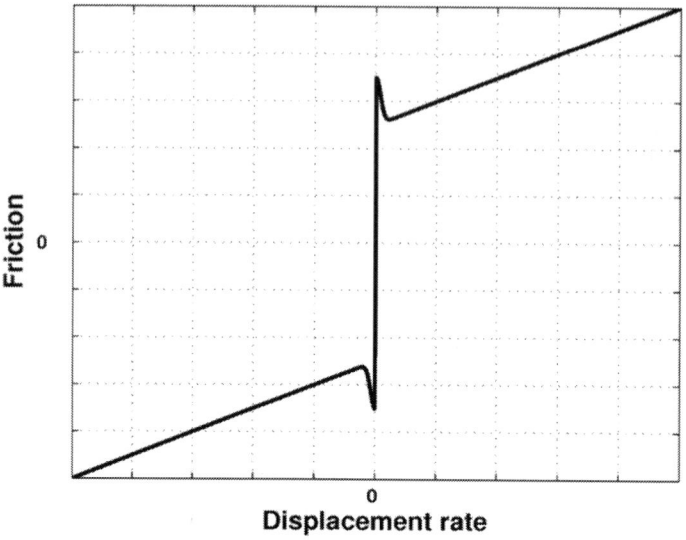

Figure 2. Friction model.Friction

In a single stage speed reduction system, a lumped flexibility model can be used if the dominant flexion appears in the gear teeth. In the following, the input gears inertia are combined with the actuators inertia Jm while the output gears inertia is lumped with the load inertia $M(q)$. As shown in [38] and the references therein, this model reduction method has been used in multi-stage reduction systems, such as planetary gears, and in multi-mass flexibility models, like harmonic drives, for instance.

2.3. Problem Statement

Given the desired trajectories qd and $q\dot{}d$, the aim is to design a control law τm which ensures that the manipulator's position q and velocity $q\dot{}$ track their desired trajectories under unknown or uncertain dynamics and in the presence of external disturbances. The proposed controller uses q, $q\dot{}$, and $\theta\dot{}$ as the system's measurable states. The manipulator's parameters, $M(q)$, $C(q, \dot{q})$, $G(q)$, J_m, τ_{fl}, τ_{fm}, τ_{dl}, τ_{dm} τfm, τdl, τdm are assumed to be unknown or uncertain.

3. TYPE-2 FLSS

A type-2 FLS is comprised of five components: fuzzifier, rule base, fuzzy inference engine, type-reducer and defuzzifier. A block diagram of a typical type-2 FLS is depicted in Figure 3.

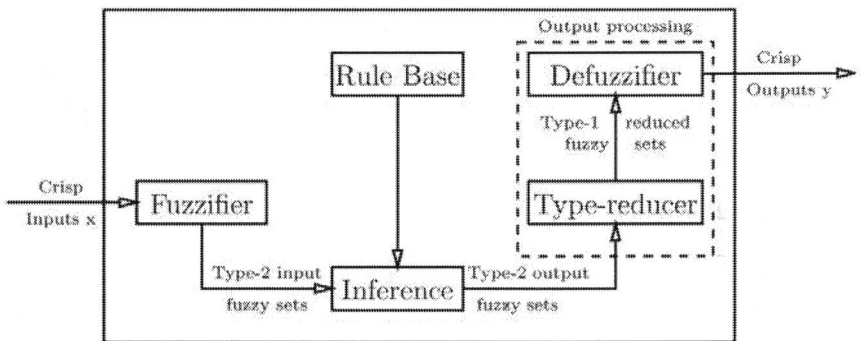

Figure 3. Block diagram of a type-2 FLS.

A type-2 fuzzy set is defined by a fuzzy membership function, where the membership value or grade for each element of this set is a fuzzy set in the interval [0,1] rather than a crisp value. As such, the footprint of uncertainty (FOU) provides type-2 FLSs with additional degrees of freedom, making the membership functions of type-2 fuzzy sets three dimensional functions. Therefore, type-2 fuzzy sets can handle more types of uncertainties with higher magnitudes using a smaller rule base than their type-1 counterparts.

A type-2 fuzzy set, denoted as $\tilde{A}$, is characterized by a type-2 membership function $\mu_{\tilde{A}}(x, u)$, where $x \in X$ and $u \in J_x \subseteq [0,1]$, *i.e.*

$$\tilde{A} = \{((x, u), \mu_{\tilde{A}}(x, u) \mid \quad \forall x \in X, \quad \forall u \in J_x \subseteq [0,1]\}$$

in which $0 \leq \mu_{\tilde{A}}(x, u) \leq 1$. For a continuous universe of discourse, $\tilde{A}$ can be expressed as

$$\tilde{A} = \int_{x \in X} \int_{u \in J_x} \mu_{\tilde{A}}(x, u)/(x, u) \quad J_x \subseteq [0, 1]\}$$

where Jx is referred to as the primary membership of x. As in type-1 fuzzy logic, discrete fuzzy sets are represented by the symbol Σ instead of $\int$. The secondary membership function associated to $x = x'$,', for a given $x' \in X$, is the type-1 membership function defined by $\mu_{\tilde{A}}(x = x', u)$, $\forall u \in J_x$. The uncertainty in the primary membership of a type-2 fuzzy set $A^\sim$ is represented by the FOU and is illustrated in Figure 4. Note that the FOU is also the union of all primary memberships.

$$FOU(\tilde{A}) = \bigcup_{x \in X} J_x$$

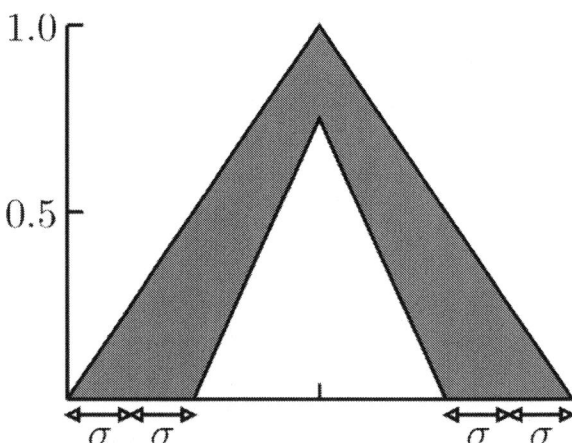

Figure 4. Type-2 fuzzy logic membership function.

The upper and lower membership functions, denoted by $\mu^{\sim\sim}A^\sim(x)$ and $\mu{-}A^\sim(x)$, respectively, are two type-1 membership functions that represent the upper and lower bounds for the footprint of uncertainty of an interval type-2 membership function $\mu A^\sim(x, u)$, respectively [56].

The structure of rules in the type-2 FLS and its inference engine is similar to those in type-1 FLS. The inference engine combines rules and provides a mapping from input type-2 fuzzy sets to output type-2 fuzzy sets. In this process, unions and intersections of type-2 sets as well as compositions of type-2 relations are used. The output of the type-2 inference engine is a type-2 set. Using extension principle, type-1 defuzzification can derive a crisp output from type-1 fuzzy set; similarly, for a higher type set as type-2, this operation derives the type-2 sets to a type-1 set. This process is the so called "type-reduction".

4. INTERVAL TYPE-2 FLSS

The computation intensity of type-2 FLSs has been behind the development of interval fuzzy sets [56,57] as it provides a simplified and efficient way to compute meet and join operations and perform type-reduction for FLSs. It distributes the uncertainty evenly among all admissible primary memberships and offers a balanced trade-off between performance and complexity. Before proceeding further, a few concepts need to be introduced. An FLS with an interval singleton type-2 fuzzifier and product or minimum t-norm satisfies the following properties [57]:

- the firing strength of the lth fuzzy rule is an interval type-1 fuzzy set defined as

$$F^l(x') \equiv \sqcap_{i=1}^p \mu_{\tilde{F}_i^l}(x_i') = \left[\underline{f}^l(x'), \overline{f}^l(x') \right] \equiv \left[\underline{f}^l, \overline{f}^l \right]$$

where

$$\underline{f}^l(x') = \mu_{\tilde{F}_1^l}(x_1') \star \cdots \star \mu_{\tilde{F}_p^l}(x_p')$$

(3)

$$\overline{f}^l(x') = \overline{\mu}_{\tilde{F}_1^l}(x_1') \star \cdots \star \overline{\mu}_{\tilde{F}_p^l}(x_p')$$

(4)

with the t-norm operator denoted by "$\star$".

- the fired output consequent set of the lth rule is a type-1 fuzzy set characterized by a membership function

$$\mu_{\tilde{B}^l}(y) = \int_{b^l \in [\underline{f}^l \star \underline{\mu}_{\tilde{G}^l}(y), \overline{f}^l \star \overline{\mu}_{\tilde{G}^l}(y)]} 1/b^l \qquad \forall y \in Y$$

with $\underline{\mu}_{\tilde{G}^l}(y)$ (y) and $\overline{\mu}_{\tilde{G}^l}(y)$ being the lower and upper membership grades of $\mu_{\tilde{G}^l}(y)$.

- if N out of a total of L fuzzy rules in the FLS fire, where $N \leq L$, then the overall aggregated output fuzzy set is defined by a type-1 membership function $\mu_{\tilde{B}}(y)$ obtained by combining the fired output consequent sets into one. In other words, $\mu_{\tilde{B}}(y) = \sqcup_{l=1}^{N}\mu_{\tilde{B}^l}(y)$, is defined in Equation (5).

The following is a brief description of the different stages of the type-2 fuzzy logic inference engine.

$$\mu_{\tilde{B}^l}(y) = \int_{b \in \left[[\underline{f}^1 \star \underline{\mu}_{\tilde{G}^1}(y)] \vee \cdots \vee [\underline{f}^N \star \underline{\mu}_{\tilde{G}^N}(y)], [\overline{f}^1 \star \overline{\mu}_{\tilde{G}^1}(y)] \vee \cdots \vee [\overline{f}^N \star \overline{\mu}_{\tilde{G}^N}(y)]\right]} 1/b^l,$$

$$1/b^l, \quad \forall y \in Y$$

$$(5)$$

4.1. Type-2 Fuzzification

In the fuzzification stage, the crisp input vector with n elements $x=(x1,...,xn)T$ in the universe of discourse $X_1 \times X_2 \times ... \times X_n$ is mapped into type-2 fuzzy sets [56,57]. The type-2 fuzzification process is schematically depicted in Figure 5. The upper and lower membership functions are computed for each point of the universe of discourse, resulting in an interval type-1 set $[\underline{f}^l, \overline{f}^l]$ for each rule l.

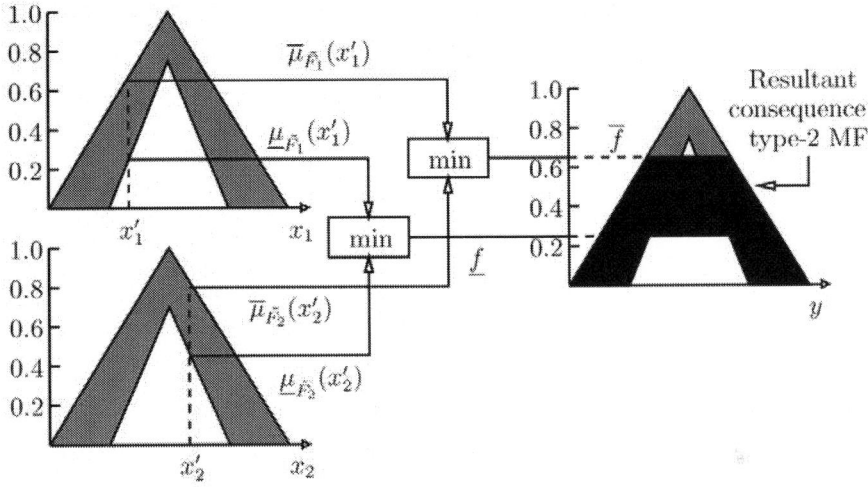

Figure 5. Interval type-2 inference process.

4.2. Type-2 Fuzzy Rule Base

Type-2 IF-THEN rules have the same structure as their type-1 counterpart. However, the antecedents and the consequents are represented by interval type-2 fuzzy sets. Thus, the *lth* rule is of the form:

$$R^l : \text{IF } x_1 \text{ is } \tilde{F}_1^l \text{ and } x_2 \text{ is } \tilde{F}_2^l \text{ and } \dots \text{ and } x_n \text{ is } \tilde{F}_n^l$$
$$\text{THEN } y_1 \text{ is } \tilde{G}_1^l \text{ and } y_2 \text{ is } \tilde{G}_2^l \text{ and } \dots \text{ and } y_m \text{ is } \tilde{G}_m^l$$

where •is type-2 fuzzy set.

4.3. Type-2 Fuzzy Inference Engine

A type-2 fuzzy inference engine provides a mapping from the input type-2 fuzzy sets to the output ones. Each rule l in the knowledge base is interpreted as a type-2 fuzzy implication that, when aggregated with the fuzzified inputs, infers a type-2 fuzzy set $B^\sim l$ such that:

$$\mu_{\tilde{B}^l}(y) = \bigsqcup_{x \in X} \left[\mu_{\tilde{A}_x}(x) \sqcap \mu_{R^l}(x, y) \right]$$

The t-norm and t-conorm used for the type-2 FLC herein are the "minimum" and "maximum" operators, respectively. These operators have been accredited in the literature for their computational efficiency and satisfactory performance.

4.4. Type Reduction

Type-reduction is an "extended version" of type-1 defuzzification methods (using the extension principle) because this operation reduces a type-2 output fuzzy set to a type-1 fuzzy set, which is called the "type-reduced set". Several type-reduction methods have been suggested in the literature, such as the center-of-sums, the height, the modified height and the center-of-sets. The calculation of type-reduced sets is performed in two stages. First, the centroids of the type-2 interval consequent sets of the fuzzy rules are computed. This is conducted ahead of time and is not part of the control cycle. In the second stage, the type-reduced sets are computed at each control cycle before being defuzzified [54].

Calculation of the Rule Consequents Centroids The centroid of the tth output fuzzy set y_k^t is a type-1 interval set determined by its left and right most points, y_{lk}^t and y_{rk}^t, respectively, which are expressed by [57]:

$$y_k^t = [y_{lk}^t, y_{rk}^t] = y(\theta_1, \ldots, \theta_z) = \frac{\sum_{z=1}^{Z} y_z \theta_z}{\sum_{z=1}^{Z} \theta_z} \tag{6}$$

Algorithm 1 describes the iterative procedures for computing y_{rk}^t and y_{lk}^t [54]. Without loss of generality, we assume that $y_z, z = 1, \ldots, Z$, are arranged in an ascending order; i.e., $y_1 \leq y_2 \leq \cdots \leq y_Z$.

Note that both algorithms are guaranteed to converge in at most Z iterations. However, as mentioned earlier, these procedures are conducted ahead of time and are not part of the control cycle. Such a property helps alleviate the heavy computational burden that is usually associated to computational intelligence-based controllers in general.

4.5. Calculation of the Type-Reduced Set

Using the centroid method, the center-of-sets type reduction reduces the resulting type-2 fuzzy sets to an interval type-1 fuzzy set $[y_{lk}^i, y_{rk}^i]$ for each rule i. The inferred interval type-1 fuzzy set is then defined by $[y_{lk}, y_{rk}]$, such as:

$$y_{lk} = \frac{\sum_{i=1}^{L} f_l^i y_{lk}^i}{\sum_{i=1}^{L} f_l^i} \tag{7}$$

$$y_{rk} = \frac{\sum_{i=1}^{L} f_r^i y_{rk}^i}{\sum_{i=1}^{L} f_r^i} \tag{8}$$

where f_l^i, f_r^i are the firing strengths corresponding to y_{lk}^i and y_{rk}^i of rule i, to minimize y_{lk}^i and maximize y_{rk}^i. Algorithms 2 and 3 reveal the iterative procedures to compute and , respectively [57]. Without loss of generality, we here assume that the pre-computed y_{lk}^i, i=1,...,L, are arranged in an ascending order; i.e., $y_{lk}^1 \leq y_{lk}^2 \leq \ldots \leq y_{lk}^L$. Hence, y_{lk} and y_{rk} can be mathematically expressed as:

$$y_{lk} = \frac{\sum_{u=1}^{Q} \overline{f}^u y_{lk}^u + \sum_{v=Q+1}^{L} \underline{f}^v y_{lk}^v}{\sum_{u=1}^{Q} \overline{f}^u + \sum_{v=Q+1}^{L} \underline{f}^v}$$

$$y_{rk} = \frac{\sum_{u=1}^{R} \underline{f}^u y_{rk}^u + \sum_{v=R+1}^{L} \overline{f}^v y_{rk}^v}{\sum_{u=1}^{R} \underline{f}^u + \sum_{v=R+1}^{L} \overline{f}^v}$$

It is worth pointing out that both procedures are proven to converge in no more than L iterations, where L is the total number of rules [57].

Algorithm 1: Computing y_{rk}^t or y_{lk}^t.

begin

  Set $\theta_z = h_z$ for $z = 1, \ldots, Z$;

  Compute $y' = y(h_1, \ldots, h_z)$ using Equation (6);

  Set Stop = False;

  while *Stop = False* **do**

    Find e, where $1 \leq e \leq Z - 1$, such that $y_e \leq y' \leq y_{e+1}$;

    if y_{rk}^t *is to be computed* **then**

      Set $\theta_z = h_z - \Delta_z$ for $z \leq e$;

      Set $\theta_z = h_z + \Delta_z$ for $z > e + 1$;

      Compute $y'' = y(h_1 - \Delta_1, \ldots, h_e - \Delta_e, h_{e+1} + \Delta_{e+1}, \ldots, h_z + \Delta_z)$ using Equation (6);

    else if y_{lk}^t *is to be computed* **then**

      Set $\theta_z = h_z + \Delta_z$ for $z \leq e$;

      Set $\theta_z = h_z - \Delta_z$ for $z > e + 1$;

      Compute $y'' = y(h_1 + \Delta_1, \ldots, h_e + \Delta_e, h_{e+1} - \Delta_{e+1}, \ldots, h_z - \Delta_z)$ using Equation (6);

    if $y'' = y'$ **then**

      Stop = True;

      if y_{rk}^t *is to be computed* **then**

        Set $y_{rk}^t = y''$ (y'' is the maximum value of $y(\theta_1, \ldots, \theta_z)$);

      else if y_{lk}^t *is to be computed* **then**

        Set $y_{lk}^t = y''$;

    else

      Set $y' = y''$;

end

Algorithm 2: Computing y_{lk}.

begin

  Set $f_r^i = (\underline{f}^i + \overline{f}^i)/2$, for $i = 1, \ldots, L$, with $\underline{f}^i$ and $\overline{f}^i$ as defined in Equations (3) and (4);

  Set $y_{lk}' = y_{lk}$;

  Set Stop = False;

  while *Stop = False* **do**

    Find Q $(1 \leq Q \leq L - 1)$ such that $y_{lk}^Q \leq y_{lk}' \leq y_{lk}^{Q+1}$;

    Compute y_{lk} as in Equation (7), using $f_l^i = \overline{f}^i$ for $i \leq Q$ and $f_l^i = \underline{f}^i$ for $i > Q$;

    Let $y_{lk}'' = y_{lk}$;

    if $y_{lk}'' = y_{lk}'$ **then**

      Stop = True;

      Set $y_{lk}'' = y_{lk}$;

    else

      Set $y_{lk}' = y_{lk}''$;

end

Algorithm 3: Computing y_{rk}.

begin

    Set $f_r^i = (\underline{f}^i + \overline{f}^i)/2$. for $i = 1, \ldots, L$. with $\underline{f}^i$ and $\overline{f}^i$ as defined in Equations (3) and (4);

    Set $y'_{rk} = y_{rk}$;

    Set Stop $=$ False;

    while *Stop $=$ False* **do**

        Find R ($1 \leq R \leq L - 1$) such that $y_{rk}^R \leq y'_{rk} \leq y_{rk}^{R+1}$;

        Compute y_{lk} as in Equation (8), using $f_r^i = \overline{f}^i$ for $i \leq R$ and $f_r^i = \underline{f}^i$ for $i > R$;

        Let $y''_{rk} = y_{rk}$;

        if $y''_{rk} = y'_{rk}$ **then**

            Stop $=$ True;

            Set $y''_{rk} = y_{rk}$;

        else

            Set $y'_{rk} = y''_{rk}$;

end

4.6. Type-2 Defuzzification

The type-reduced set $Ycos(X)k$ determined by its left most and right most points, ylk and yrk, respectively, is defuzzified using the interval set average formula to get a crisp output value. As such, the defuzzified crisp output for each output k is formulated as [56]:

$$Y_k(x) = \frac{y_{lk} + y_{rk}}{2}$$

In this work, we are concerned with adaptive type-2 FLC, and hence, Algorithms 1–3 will be replaced with the adaptive mechanism explained in the next section. As such, we overcome the heavy computational burden traditionally associated to computational intelligence-based controllers.

5. CONTROL STRATEGY

Let $\Delta q = qd - q$ and $\Delta \theta = \theta d - \theta$ denote the links' and actuators' position errors, respectively, with θd being the unknown desired time-dependent motor position vector. The control strategy is based on the design of an adaptive controller that not only leads to a precise tracking of the system's nominal desired signals but also improves the motors' internal stability. Should the motors' desired position θd have been available, the control strategy would be based on tracking Δq and $\Delta\theta$ to zero. Since that is not the case, we define the following compounded velocity error signal:

$$\Delta\dot{e}_r = \dot{q}_d - \left(\lambda\dot{q} + (1-\lambda)\frac{1}{r}\dot{\theta} \right)$$

(9)

for a diagonal matrix $\lambda=\text{diag}\ (\lambda 1,\lambda 2,...,\lambda n)$ with $\lambda i\in\ [0,1]$, $i=1,...,n$. The feedback gain λ is introduced to provide a trade-off between the link tracking performance and internal stability, due to the high nonlinear coupling between the two. Note how a choice of $\lambda i=1$, $\forall i\in\{1,...,n\}$, completely annihilates the motor's internal stability factor. The fuzzy control strategy is based on a human operator experience to interpret a situation and initiate its control action. A block diagram for the fuzzy controller is illustrated in Figure 6. Given the desired control signals qd and $q\dot{}d$, the link's position error Δq and the compounded velocity error $\Delta\epsilon\dot{}r$ are computed. The FLC takes these two inputs and provides a control action τm that is proportional to the input values. These signals are quantized into 5 levels represented by a set of linguistic variables: Negative Large (NL), Negative Small (NS), Zero (Z), Positive Small (PS), and Positive Large (PL). In this study, triangular membership functions are used, mainly due to their high computational and performance efficiencies [58]. To assess the performance of both types of controllers, the proposed fuzzy controller is implemented in two different ways: the first is based on a type-1 fuzzy control scheme while the second is based on a type-2. The input membership functions adopted by both types of control systems are shown in Figure 7. In order to perform a fair comparison between type-1 and type-2 FLCs, the type-2 membership functions have been designed such that they are very similar to those of type-1 and have small blurbs in their FOUs. The fuzzy rules of the two control techniques are the same, they were chosen heuristically and can be refined by an expert (see Table 1). These rules are based on three hypotheses: (i) when the input signals are far from their respective nominal zero-valued surfaces, then the FLC's output assumes a high value; (ii) when the inputs are approaching the nominal zero-valued surfaces, the output is adjusted to a smaller value for a smoother approach; (iii) once the inputs are on the nominal zero-valued surfaces, then the output is set to zero. This way, the FLC forces the link position error Δq and the compounded velocity error $\Delta \acute{\epsilon} r$ to approach zero. It is worth mentioning that an empirical study was conducted beforehand to tune the input membership functions. The center of area method is used for defuzzification.

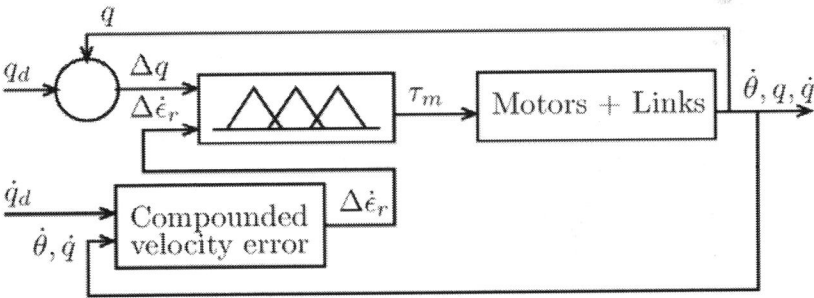

Figure 6. Block diagram of the proposed fuzzy control scheme.

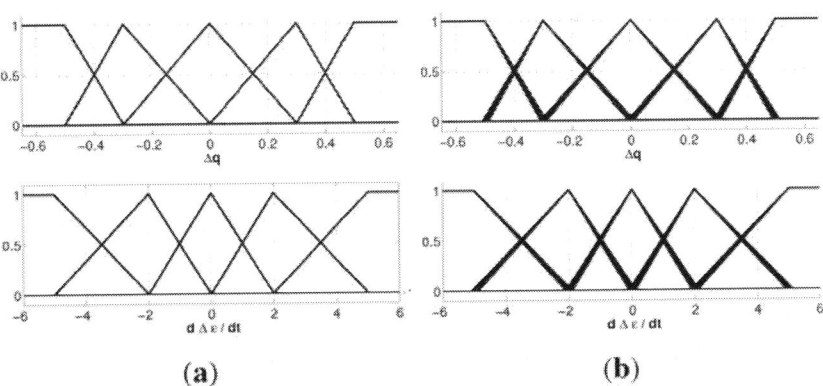

Figure 7. Fuzzy membership functions: (**a**) type-1; and (**b**) type-2.

Table 1. Fuzzy rules for type-1 and type-2 FLCs.

$\Delta \dot{\epsilon}_r$	Δq				
	NL	**NS**	**Z**	**PS**	**PL**
PL	Z	PL	PL	PL	PL
PS	NS	Z	PS	PS	PL
Z	NL	NS	Z	PS	PL
NS	NL	NS	NS	Z	PS
NL	NL	NL	NL	NL	Z

5.1. Adaptive Type-2 FLC

The adaptive type-2 FLC structure is depicted in Figure 8. It consists of four layers. Input nodes and type-2 fuzzification nodes are shown in layer 1 and layer 2, respectively, forming the antecedent part of the fuzzy rules. Consequent parts are represented by layers 3 and 4, which are constructed with fuzzy rule nodes and output nodes. They are linked by interval weighting factors [wmlzwmrz].

The adaptive type-2 FLC's output can be written as:

$$Y = \Phi^T w + \epsilon = \Phi^T \hat{w}$$

(10)

where $\in \mathbb{R}^{z \times m}$ is a weight matrix and Φ is a m-dimensional vector of known functions (regressor).

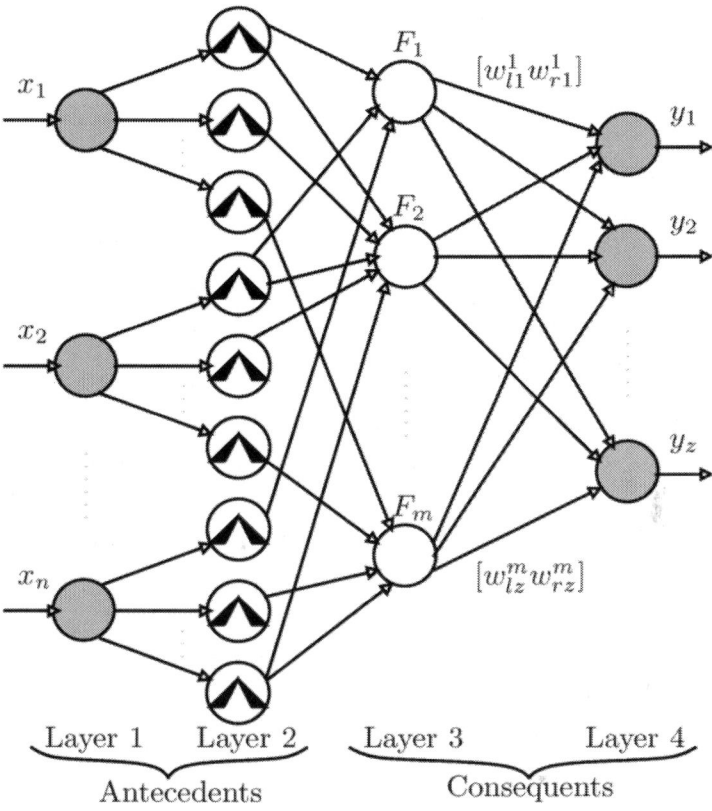

Figure 8. Adaptive Type-2 fuzzy logic control structure.

Define the following signals:

$$s = \Delta\dot{e}_r + \Psi\Delta q = \dot{q}_r - \Lambda\dot{q} - (1-\Lambda)\frac{1}{r}\dot{\theta} \tag{11}$$

$$\dot{q}_r = \dot{q}_d + \Psi\Delta q \tag{12}$$

where $\Psi = [\psi_1, \psi_2, \ldots, \psi_n]$ with ψ_i being a positive constant, $i=1,\ldots,n$.

Recall Euler–Lagrange formulation (1):

$$M(q)\ddot{q} + C(q,\dot{q})\dot{q} + G(q) + \tau_{fl} + \tau_{dl} - \tau_t = 0$$
$$J_m\ddot{\theta} + \tau_{fm} + \tau_{dm} - \tau_m = -\frac{1}{r}\tau_t$$

where $q, \dot{q}, \ddot{q}, \ddot{\theta} \in \mathbb{R}^n$. Substituting for τt, and letting the stiffness constant K tend to infinity (singular perturbation), we obtain the following rigid model [16]:

$$M_t(q)\ddot{q} + C(q,\dot{q})\dot{q} + G(q) + \tau_{Frd} = \tau_m \tag{13}$$

where,

$$M_t(q) = J_m r + M(q)$$
$$\tau_{Frd} = \tau_{fl} + \tau_{dl} + \tau_{fm}r + \tau_{dm}r + (1-r)\tau_m$$

Using the linear in parameters property of the manipulator dynamics, we can approximate the model with a linear regression:

$$M_t(q)\ddot{q}_r + C(q,\dot{q})\dot{q}_r + G(q) + \tau_{Frd} = \Phi(\ddot{q}_r, \dot{q}_r, \dot{q}, q)^T w$$

The control law is:

$$\tau_m = \Phi^T \hat{w} + K_D s \tag{14}$$

where KD is a positive diagonal matrix gain and the sign $\hat{\bullet}$ denotes the parameter estimate vector.

Theorem 1 *Consider the nonlinear system in Equation (1) with reference signal (11) and control law (14). The adaptive control law is asymptotically stable and the tracking error converges to zero with the following adaptation law:*

$$\dot{\hat{w}} = -\Gamma\,\Phi\,s$$

where $\Gamma=diag(\gamma 1,\gamma 2,...,\gamma n)$ and γi is a positive constant, $i=1,...,n$.

Proof: Taking the derivative of the error signal s yields

$$\dot{s} = \ddot{q}_r - \Lambda\dot{q} - (1 - \Lambda)\frac{1}{r}\ddot{\theta}$$

Let $K\to\infty$, i.e., displacement $(\theta-q)\to 0$. We get:

$$\dot{s} = \ddot{q}_r - \ddot{q}$$

$$M_t(q)\dot{s} = M_t(q)\ddot{q}_r - M_t(q)\ddot{q}$$

Substituting Mq'' from Equation (13):

$$M_t(q)\dot{s} = M_t(q)\ddot{q}_r + C(q,\dot{q})\dot{q}_r + G(q) + \tau_{Frd} - C(q,\dot{q})s - \tau_m$$

The linear in parameters property yields

$$M_t(q)\dot{s} = \Phi^T w - C(q,\dot{q})s - \tau_m$$

$$(15)$$

Set

$$\tau_m = \Phi^T \hat{w} + K_D s$$

$$(16)$$

Equation (15) becomes

$$M_t(q)\dot{s} = \Phi^T \tilde{w} - C(q, \dot{q})s - K_D s$$

$$(17)$$

where $\tilde{w} = w - \hat{w}$.

Choose the following Lyapunov candidate:

$$V = \frac{1}{2} \{ s^T M_t(q)s + \tilde{w}^T \Gamma^{-1} \tilde{w} \}$$

Taking the derivative of V:

$$\dot{V} = s^T M_t(q)\dot{s} + \frac{1}{2} s^T \dot{M}_t(q)s + \tilde{w}^T \Gamma^{-1} \dot{\tilde{w}}$$

Substituting for $Mt(q)\dot{s}$:

$$\dot{V} = s^T \Phi^T \tilde{w} + s^T \{\frac{1}{2}(\dot{M}_t(q) - 2C(q, \dot{q}))\}s + \tilde{w}^T \Gamma^{-1}\dot{\hat{w}} - s^T K_D s$$

$$s^T \{(\dot{M}_t(q) - 2C(q, \dot{q}))\}s = 0 \text{ due to the skew-symmetry property.}$$

Hence,

$$\dot{V} = s^T \Phi^T \tilde{w} + \tilde{w}^T \Gamma^{-1}\dot{\hat{w}} - s^T K_D s$$

Setting the adaptation law as

$$\dot{\hat{w}} = -\Gamma \, \Phi \, s$$

leads to

$$\dot{V} = -s^T K_D \, s \leq 0$$

Therefore V, and so s, $\tilde{w}$ and $\hat{w}$, are bounded and converge to finite values. It follows from Equation (16) that τm is bounded, which implies that all the terms in Equation (17), including $s^{\cdot}$, are bounded. Thus, $\ddot{V} = -2s^T K_D \dot{s}$ is also bounded. Hence, from Barbalet's Lemma, it implies that $\lim t \to \infty V^{\cdot} = 0$. Therefore, $\lim t \to \infty s = 0$.

6. SIMULATION RESULTS AND DISCUSSION

6.1. Simulation Setup

To demonstrate the performance of the proposed controller, two numerical simulations are carried out on a single link flexible-joint manipulator. Table 2 summarizes the manipulator's physical parameters along with their respective values. The stiffness coefficient and gear ratio are set to be $K=7$ N·m/rad and $r=1$, respectively. The link's mass and length are taken as $m=0.21$ Kg and $l=0.3$ m, respectively. The manipulator's dynamics in terms of its physical parameters is defined by: $M(q)=I$, $C(q,q^{\cdot})=0$, and $G(q)=mgl\sin(q)$, where $g=9.8$ m/s2 is the gravitational constant and I is the link's rotational inertia given in Table 2. The nonlinear friction model described in Section 2 is considered to model the actuator's and load's mechanical frictions, τfm and τfl, respectively.

Table 2. Manipulators physical parameters.

Parameter	Link	Motor
rotational inertia (kg·m^2)	$I = 5.05 \times 10^{-2}$	$J_m = 4 \times 10^{-3}$
viscous friction coefficient (N·m·s/rad)	$F_{vl} = 4 \times 10^{-3}$	$F_{vm} = 3 \times 10^{-3}$
Coulomb friction coefficient (N·m)	$F_{cl} = 1 \times 10^{-2}$	$F_{cm} = 4 \times 10^{-3}$
static friction coefficient (N·m)	$F_{sl} = 2 \times 10^{-3}$	$F_{sm} = 2 \times 10^{-3}$
static friction decreasing rate (rad/s)	$\eta_{sl} = 7 \times 10^{-2}$	$\eta_{sm} = 5 \times 10^{-2}$

The manipulator's desired position trajectory is taken as the step response of a critically damped second order system with a natural frequency of 3 rad/s, as shown in Figure 9. The feedback gain λ in Equation (9) is set to 0.8. The control structure scheme and the system's model are implemented in SimulinkTM while the fuzzy control engines are programmed in C. Both controllers are set to operate at a bandwidth of 100 Hz.

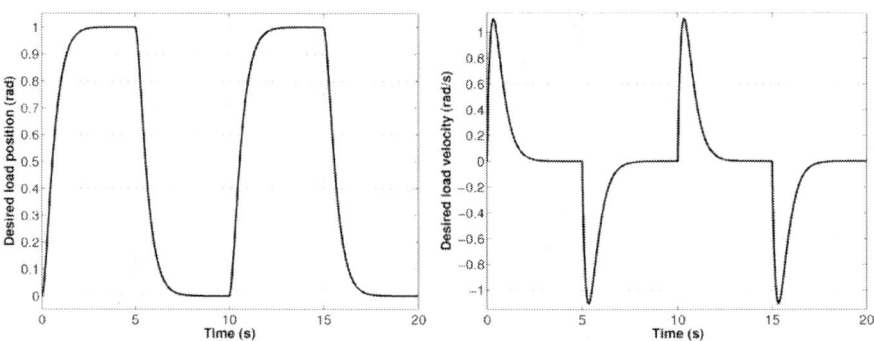

Figure 9. Manipulator's position and velocity reference signals.

6.2. Numerical Simulations and Results

Two simulations are carried out to highlight the proposed type-2 FLC as opposed to type-1 in tolerating a higher degree of parametric and modeling uncertainties. In both simulations, the system's dynamics is assumed to be a priori unknown. For each simulation, the system's response is studied taking into account the manipulator's position and velocity errors, the joint's internal stability, and the controller's output torque, τm.

The first simulation is meant to study the controller's ability to sustain various types and magnitudes of load uncertainties. For this purpose, the load's inertia and the link's mass are both doubled abruptly at 5 s and returned back to their original values at 15 s of the simulation. The results are shown in Figure 10. A slight increase is noticed in the manipulator's position and velocity errors due to a heavier load. However, the error signals and the controller's output under type-1 FLC are fairly fluctuating as opposed to a smooth and steady convergence behavior with the type-2 FLC. It is quite important to notice here the degradation in the actuator's internal stability (Figure 10(e)) under type-1 FLC despite the settling of the load's velocity. The superiority of the adaptive type-2 FLC in compensating for such a type of uncertainty is manifested with a better load position, speed accuracy and control effort performance over its type-1 counterpart.

In the second simulation, the elastic joint's stiffness coefficient is changed abruptly to $K=5$ N·m/rad at 5 s and returned back to its original value ($K=7$ N·m/rad) at 15 s of the simulation. The controller's performance under such conditions is revealed in Figure 11. As in simulation 1, adaptive type-1 FLC is able to maintain bounded error signals but fails to make them converge smoothly. This is especially clear with the load's position and velocity errors (Figure 11(a) and Figure 11(c)) and the motor's internal stability (Figure 11(e)). On the other hand, it is clear that the adaptive type-2 FLC does indeed maintain these signals smoother and with less control effort than its type-1 counterpart.

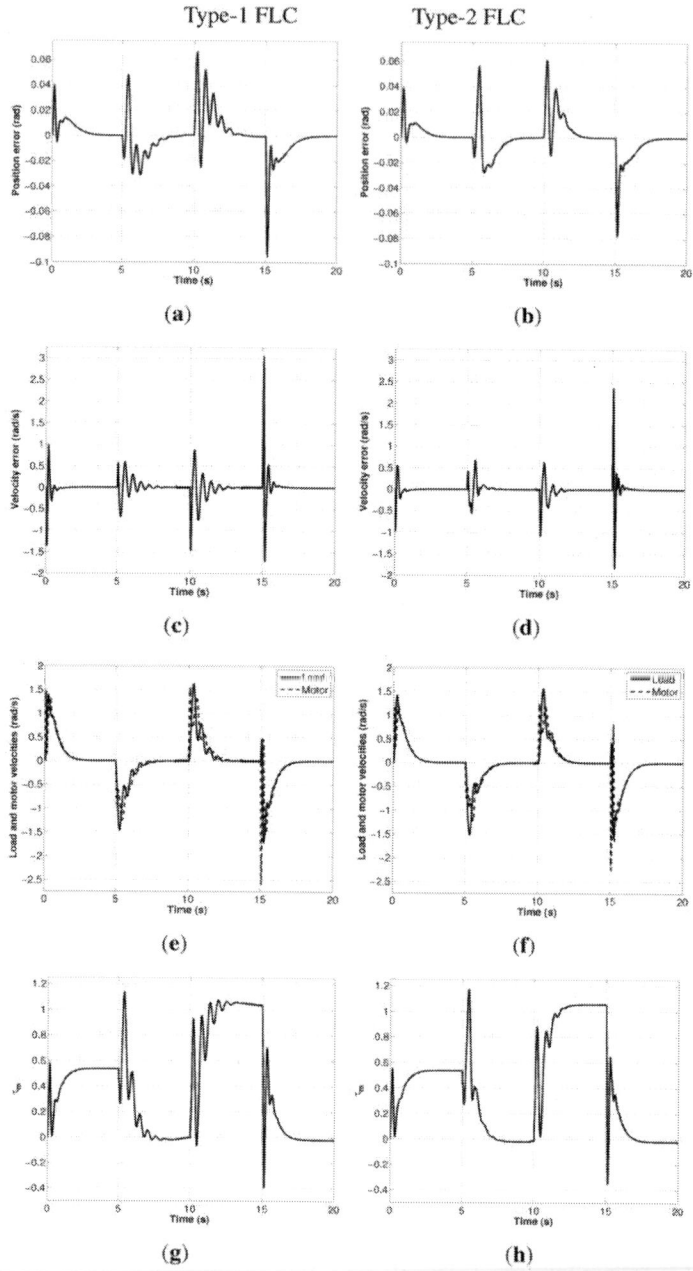

Figure 10. System's response to varying load's mass and inertia: (**a**), (**b**) manipulator's position error; (**c**), (**d**) manipulator's velocity error; (**e**), (**f**) motor's velocity vs. manipulator's velocity; and (**g**), (**h**) controller's output torque (τm).

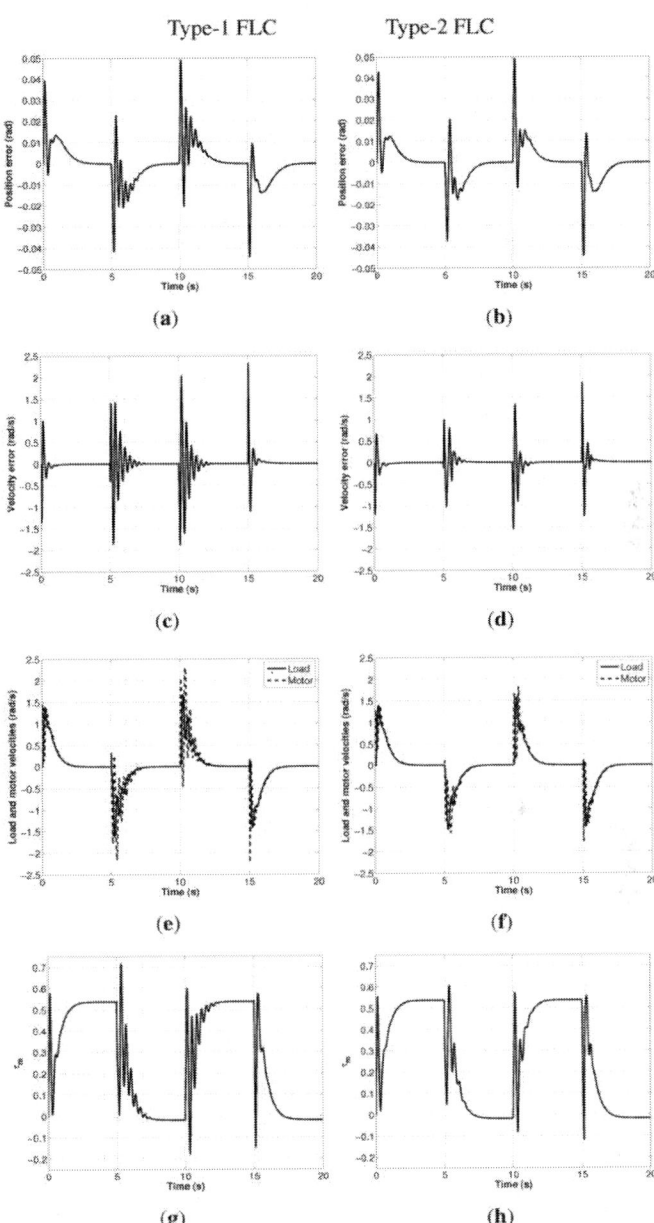

Figure 11. System's response to varying stiffness coefficient: (**a**), (**b**) manipulator's position error; (**c**), (**d**) manipulator's velocity error; (**e**), (**f**) motor's velocity vs. manipulator's velocity; and (**g**), (**h**) controller's output torque (τm).

7. CONCLUSIONS

In this paper, an adaptive type-2 FLC with a Mamdani inference engine has been proposed for flexible-joint manipulators in the presence of dynamical modeling and parametric uncertainties of various magnitudes. The control strategy takes into account the actuators' relative stability by introducing a trade-off criterion between the actuators' internal stability and the links' position. A comparison of the proposed controller and its type-1 counterpart is performed under similar operating conditions. The simulations show the superiority of the adaptive type-2 FLC in compensating for high magnitude of uncertainties, which confirms the theoretical credentials associated to type-2 FLSs.

ACKNOWLEDGEMENTS

The authors would like to acknowledge Natural Sciences and Engineering Research Council of Canada (NSERC) for partially supporting this work through the Discovery program.

CONFLICT OF INTEREST

The authors declare no conflict of interest.

REFERENCES

1. Olsson, H.; Astrom, K.; de Wit, C.C.; Gafvert, M.; Lischinsky, P. Friction models and friction compensation. Eur. J. Control **1998**, 4, 176–195.

2. Seidl, D.R.; Lam, S.L.; Putman, J.A.; Lorenz, R.D. Neural network compensation of gear backlash hysteresis in position-controlled mechanisms. IEEE Trans. Ind. Appl. **1995**, 31, 1475–1483.

3. Sweet, L.; Good, M. Redefinition of the robot motion-control problem. Control Syst. Mag. IEEE **1985**, 5, 18–25.

4. Armstrong, B.; de Wit, C.C. Friction modeling and compensation. Control Handb. **1996**, 77, 1369–1382.

5. Ghorbel, F.; Hung, J.; Spong, M. Adaptive control of flexible-joint manipulators. Control Syst. Mag. IEEE **1989**, 9, 9–13.

6. Seidl, D.; Lam, S.L.; Putman, J.; Lorenz, R. Neural network compensation of gear backlash hysteresis in position-controlled mechanisms. IEEE Trans. Ind. Appl. **1995**, 31, 1475–1483.

7. Seidl, D.; Reineking, T.; Lorenz, R. Neural Network Compensation of Gear Backlash Hysteresis in Position-Controlled Mechanisms. In Proceedings of Conference Record of the IEEE Industry Applications Society Annual Meeting, Houston, TX, USA, October 1992; Volume 2, pp. 1937–1944.

8. Benosman, M.; Vey, G.L. Control of flexible manipulators: A survey. Robotica **2004**, 22, 533–545.

9. Chaoui, H.; Gueaieb, W. Type-2 fuzzy logic control of a flexible-joint manipulator. J. Intell. Robot. Syst. **2008**, 51, 159–186.

10. Chaoui, H.; Sicard, P.; Gueaieb, W. ANN-based adaptive control of robotic manipulators with friction and joint elasticity. IEEE Trans. Ind. Electron. **2009**, 56, 3174–3187.

11. Luca, A.D.; Isidori, A.; Nicolo, F. Control of Robot Arm with Elastic Joints via Nonlinear Dynamic Feedback. In Proceedings of the IEEE Conference on Decision and Control Including the Symposium on Adaptive Processes, Ft. Lauderdale, FL, USA, 11–13 December 1985; pp. 1671–1679.

12. Khorasani, K. Nonlinear feedback control of flexible joint manipulators: A single link case study. IEEE Trans. Autom. Control **1990**, 35, 1145–1149.

13. De Wit, C. Robust control for servo-mechanisms under inexact friction compensation. Automatica **1993**, 29, 757–761.

14. Ott, C.; Albu-Schaffer, A.; Hirzinger, G. Comparison of Adaptive and Nonadaptive Tracking Control Laws for a Flexible Joint Manipulator. In Proceedings of the IEEE International Conference on Intelligent Robots

and Systems, Lausanne, Switzerland, 30 September–4 October 2002; Volume 2, pp. 2018–2024.

15. Al-Ashoor, R.; Patel, R.; Khorasani, K. Robust adaptive controller design and stability analysis for flexible-joint manipulators. IEEE Trans. Syst. Man Cybern. **1993**, 23, 589–602.

16. Ghorbel, F.; Spong, M.W. Adaptive Integral Manifold Control of Flexible Joint Robot Manipulators. In Proceedings of the IEEE International Conference on Robotics and Automation, Nice, France, 12–14 May 1992; Volume 1, pp. 707–714.

17. Spong, M.W. Modeling AND control of elastic joint robots. J. Dyn. Syst. Meas. Control **1987**, 109, 310–319.

18. Ge, S.S.; Postlethwaite, I. Adaptive neural network controller design for flexible joint robots using singular perturbation technique. Trans. Inst. Meas. Control **1995**, 17, 120–131.

19. Taghirad, H.; Khosravi, M. Design and Simulation of Robust Composite Controllers for Flexible Joint Robots. In Proceedings of the IEEE International Conference on Robotics and Automation, Taipei, Taiwan, 14–19 September 2003; Volume 3, pp. 3108–3113.

20. Huang, L.; Ge, S.; Lee, T. Adaptive Position/force Control of an Uncertain Constrained Flexible Joint Robots-Singular Perturbation Approach. In Proceedings of the SICE Annual Conference, Sapporo, Japan, 4–6 August 2004; pp. 1693–1698.

21. Spong, M. Modeling and control of elastic joint robots. J. Dyn. Syst. Meas. Control **1987**, 109, 310–319.

22. Khorasani, K. Adaptive control of flexible joint robots. IEEE Trans. Robot. Autom. **1992**, 8, 250–267.

23. Ott, C.; Albu-Schaffer, A.; Kugi, A.; Stramigioli, S.; Hirzinger, G. A Passivity Based Cartesian Impedance Controller for Flexible Joint Robots-Part I: Torque Feedback and Gravity Compensation. In Proceedings of the IEEE International Conference on Robotics and

Automation, New Orleans, LA, USA, 26 April–1 May 2004; pp. 2659–2665.

24. Tian, L.; Goldenberg, A. Robust Adaptive Control of Flexible Joint Robots with Joint Torque Feedback. In Proceedings of the IEEE International Conference on Robotics and Automation, Nagoya, Japan, May 1995; Volume 1, pp. 1229–1234.

25. Sun, T.; Pei, H.; Pan, Y.; Zhou, H.; Zhang, C. Neural network-based sliding mode adaptive control for robot manipulators. Neurocomputing **2011**, 74, 2377–2384.

26. Li, Y.; Tong, S.; Li, T. Adaptive fuzzy output feedback control for a single-link flexible robot manipulator driven DC motor via backstepping. Nonlinear Anal. R. World Appl. **2013**, 14, 483–494.

27. Yen, H.M.; Li, T.H.S.; Chang, Y.C. Adaptive neural network based tracking control for electrically driven flexible-joint robots without velocity measurements. Comput. Math. Appl. **2012**, 64, 1022–1032.

28. Lopez-Araujo, D.J.; Zavala-Rio, A.; Santibanez, V.; Reyes, F. Output-feedback adaptive control for the global regulation of robot manipulators with bounded inputs. Int. J. Control Autom. Syst. **2013**, 11, 816–822.

29. Fateh, M.M. Nonlinear control of electrical flexible-joint robots. Nonlinear Dyn. **2011**, 67, 2549–2559.

30. Fateh, M.M.; Khorashadizadeh, S. Robust control of electrically driven robots by adaptive fuzzy estimation of uncertainty. Nonlinear Dyn. **2012**, 69, 1465–1477.

31. Islam, S.; Liu, P.X. Robust sliding mode control for robot manipulators. IEEE Trans. Ind. Electron. **2011**, 58, 2444–2453.

32. Corradini, M.L.; Fossi, V.; Giantomassi, A.; Ippoliti, G.; Longhi, S.; Orlando, G. Discrete time sliding mode control of robotic manipulators: Development and experimental validation. Control Eng. Pract. **2012**, 20, 816–822.

33. Islam, S.; Liu, P.X. Robust adaptive fuzzy output feedback control system for robot manipulators. ASME/IEEE Trans. Mechatron. **2011**, 16, 288–296.

34. Karray, F.; Gueaieb, W.; Al-Sharhan, S. The hierarchical expert tuning of PID controllers using tools of soft computing. IEEE Trans. Syst. Man Cybern. **2002**, 32, 77–90.

35. De Silva, C.W. Intelligent Control Fuzzy Logic Applications; CRC Press: Boca Raton, FL, USA, 1995.

36. Kim, E. Output feedback tracking control of robot manipulators with model uncertainty via adaptive fuzzy logic. IEEE Trans. Fuzzy Syst. **2004**, 12, 368–378.

37. Gueaieb, W.; Karray, F.; Al-Sharhan, S. A robust adaptive fuzzy position/force control scheme for cooperative manipulators. IEEE Trans. Control Syst. Technol **2003**, 11, 516–528.

38. Chaoui, H.; Sicard, P.; Lakhsasi, A.; Schwartz, H. Neural Network Based Model Reference Adaptive Control Structure for a Flexible Joint with Hard Nonlinearities. In Proceedings of the IEEE International Symposium on Industrial Electronics, Ajaccio, France, 4–7 May 2004; Volume 1, pp. 271–276.

39. Chaoui, H.; Sicard, P.; Lakhsasi, A. Reference Model Supervisory Loop for Neural Network Based Adaptive Control of a Flexible Joint with Hard Nonlinearities. In Proceedings of the IEEE Canadian Conference on Electrical and Computer Engineering, Niagara Falls, ON, Canada, 2–5 May 2004; Volume 4, pp. 2029–2034.

40. Chaoui, H.; Gueaieb, W.; Yagoub, M.; Sicard, P. Hybrid Neural Fuzzy Sliding Mode Control of Flexible-Joint Manipulators with Unknown Dynamics. In Proceedings of the 32nd Annual Conference of the IEEE Industrial Electronics Society (IECON-2006), Paris, France, 6–10 November 2006; pp. 4082–4087.

41. Chaoui, H.; Gueaieb, W.; Yagoub, M.C. Artificial Neural Network Control of a Flexible-Joint Manipulator under Unstructured Dynamic Uncertainties. In Proceedings of the IEEE International Workshop on Robotic and Sensors Environments, Ottawa, ON, Canada, 12–13 October 2007.

42. Hui, D.; Fuchun, S.; Zengqi, S. Observer-based adaptive controller design of flexible manipulators using time-delay neuro-fuzzy networks. J. Intell. Robot. Syst. Theory Appl. **2002**, 34, 453–466.

43. Subudhi, B.; Morris, A. Singular perturbation based neuro-H infinity control scheme for a manipulator with flexible links and joints. Robotica **2006**, 24, 151–161.

44. Park, C.W. Robust stable fuzzy control via fuzzy modeling and feedback linearization with its applications to controlling uncertain single-link flexible joint manipulators. J. Intell. Robot. Syst. Theory Appl. **2004**, 39, 131–147.

45. Li, T.H.S.; Huang, Y.C. MIMO adaptive fuzzy terminal sliding-mode controller for robotic manipulators. Inform. Sci. **2011**, 180, 4641–4660.

46. Cazarez-Castro, N.R.; Aguilar, L.T.; Castillo, O. Designing type-1 and type-2 fuzzy logic controllers via fuzzy lyapunov synthesis for nonsmooth mechanical systems. Eng. Appl. Artif. Intell. **2012**, 25, 971–979.

47. Castillo, O.; Melin, P. Optimization of type-2 fuzzy systems based on bio-inspired methods: A concise review. Inform. Sci. **2012**, 205, 1–19.

48. Castillo, O.; Martinez-Marroquin, R.; Melin, P.; Valdez, F.; Soria, J. Comparative study of bio-inspired algorithms applied to the optimization of type-1 and type-2 fuzzy controllers for an autonomous mobile robot. Eng. Appl. Artif. Intell. **2012**, 192, 19–38.

49. Melin, P.; Astudillo, L.; Castillo, O.; Valdez, F.; Garcia, M. Optimal design of type-2 and type-1 fuzzy tracking controllers for autonomous

mobile robots under perturbed torques using a new chemical optimization paradigm. Expert Syst. Appl. **2013**, 40, 3185–3195.

50. Feng, G. A survey on analysis and design of model-based fuzzy control systems. IEEE Trans. Fuzzy Sets Syst. **2006**, 14, 676–697.

51. Chen, C.S. Supervisory adaptive tracking control of robot manipulators using interval type-2 TSK fuzzy logic system. IET Control Theory Appl. **2011**, 5, 1796–1807.

52. Linda, O.; Manic, M. Uncertainty-robust design of interval type-2 fuzzy logic controller for delta parallel robot. IEEE Trans. Ind. Inf. **2011**, 57, 661–670.

53. Biglarbegian, M.; Melek, W.W.; Mendel, J.M. Design of novel interval type-2 fuzzy controllers for modular and reconfigurable robots: Theory and experiments. IEEE Trans. Ind. Electron. **2011**, 58, 1371–1384.

54. Hagras, H.A. A hierarchical type-2 fuzzy logic control architecture for autonomous mobile robots. IEEE Trans. Fuzzy Syst. **2004**, 12, 524–539.

55. Sicard, P. Trajectory Tracking of Flexible Joint Manipulators with Passivity Based Controller. Ph.D. Thesis, Rensselaer Polytechnic Institute, Troy, NY, USA, 1993.

56. Liang, Q.; Mendel, J.M. Interval type-2 fuzzy logic systems: Theory and design. IEEE Trans. Fuzzy Syst. **2000**, 8, 535–550.

57. Mendel, J.M. Uncertain Rule-Based Fuzzy Logic Systems: Introduction and New Directions; Prentice-Hall: Upper Saddle River, NJ, USA, 2001.

58. Pedryez, W. Why triangular membership functions? Fuzzy Sets Syst. **1994**, 64, 21–30.

CHAPTER 2

Design of Fuzzy Controller for Robot Manipulators Using Bacterial Foraging Optimization Algorithm

Mickael Aghajarian[1], Kourosh Kiani[1], Mohammad Mehdi Fateh[2]

[1]*Department of Electrical & Computer Engineering, Semnan University, Semnan, Iran;* [2]*Department of Electrical & Robotic Engineering, Shahrood University of Technology, Shahrood, Iran.*

ABSTRACT

Trial and error method can be used to find a suitable design of a fuzzy controller. However, there are many options including fuzzy rules, Membership Functions (MFs) and scaling factors to achieve a desired performance. An optimization algorithm facilitates this process and finds an optimal design to provide a desired performance. This paper presents a novel application of the Bacterial Foraging Optimization algorithm (BFO) to design a fuzzy controller for tracking control of a robot manipulator driven by permanent magnet DC motors. We use efficiently the BFO algorithm to form the rule base and MFs. The BFO algorithm is compared with a Particle Swarm Optimization algorithm (PSO). Performance of the controller in the joint space and in the Cartesian space is evaluated. Simulation results show superiority of the BFO algorithm to the PSO algorithm.

KEYWORDS

BFO Algorithm; PSO Algorithm; Fuzzy Control; Robot Manipulator; Tracking Control

1. INTRODUCTION

A wide variety of control strategies were proposed to control robot manipulators. PID controls are certainly the most widely adopted control strategy in industry because of its simple structure and robust performance in a wide range of operating conditions. Although PID control offers the simplest and yet most efficient solution to many real world control problems [1], optimally tuning gain is quite difficult [2]. Alternatively, fuzzy control as a modelfree approach is simply designed to control complicated systems [3]. To form fuzzy rules, an exact knowledge of model is not required. Fuzzy controller is an intelligent controller using linguistic fuzzy rules to include information from experts. Consequently, fuzzy control of robot manipulators has attracted a great deal of researches to overcome uncertainty, nonlinearity and coupling by providing a model free control [4]. To design a Fuzzy Logic Controller (FLC), a major task is to determine fuzzy rules, Membership Functions (MFs) and scaling factors. Therefore, the controller is tuned until a desired performance is achieved. The evolutionary algorithms such as Bacterial Foraging Optimization (BFO), Particle Swarm Optimization (PSO), Genetic algorithm (GA), and Simulated Annealing (SA) are getting popular because of their abilities to find the global minima in both continuous and non-continuous domains.

Since a foraging organism takes a necessary action to maximize the energy per unit time under considering all the constraints such as sensing and cognitive capabilities, natural foraging strategy can be applied to real-world optimization problems. Based on such evolutionary idea, Passino proposed BFO as an optimization algorithm [5]. BFO algorithm is a new evolutionary computation technique, which also includes powerful optimization techniques like PSO [6] and ant colony optimization [7]. To improve BFO search performance, several researchers have extended the basic BFO to deal with multi-modal and high dimensional functions [8-10]. BFO algorithm has also been combined with other evolutionary algorithms [11] in order to reduce the convergence time and enhance the accuracy. Over certain real-world optimization problems, BFO has been reported to outperform many powerful optimization algorithms like GA [12] and PSO algorithms [13].

The PSO algorithm proposed by Kennedy and Eberhart [6], has proved to be very effective for solving complex optimization problems. The underlying motivation for the development of PSO algorithm was social behavior of animals such as bird flocking and fish schooling. Generally, PSO is characterized as a simple concept, easy to implement, and computationally efficient. Unlike the other heuristic techniques, PSO has a flexible and well-balanced mechanism to enhance the global and local exploration abilities [14].

Trial and error method is a major task to find a suitable design of a fuzzy controller. We may use an optimization algorithm to achieve an optimal design. This paper presents a BFO algorithm to design a fuzzy PID controller for trajectory tracking control of a robot manipulator driven by permanent magnet DC motors. Performance of the BFO algorithm is compared with a PSO algorithm in terms of Integral Time Absolute Error (ITAE) in the joint space and Integral Square Error (ISE) in the Cartesian space. This paper is organized as follows: Section 2 introduces dynamics of the robotic system. Section 3 designs a fuzzy PD + I controller. Section 4 applies the BFO and PSO algorithms for tuning the fuzzy PD + I controller. Section 5 presents simulation results and Finally Section 6 concludes the paper.

2. MANIPULATOR DYNAMICS

Robust control of robot manipulators is difficult because of complexity robot dynamics. The dynamics of an n-link robotic manipulator driven by permanent magnet dc motors is characterized by a set of highly nonlinear and strongly coupled second order differential equation [15] as

$$RK_m^{-1}\left(J_m r^{-1} + rD(q)\right)\ddot{q}$$
$$+\left(RK_m^{-1}B_m r^{-1} + RK_m^{-1}rC(q,\dot{q}) + K_b r^{-1}\right)\dot{q}$$
$$+RK_m^{-1}rg(q) = V$$

$$(1)$$

where $q \in R^n$ is a vector of generalized joint positions, $D(q) \in R^{n \times n}$ is the inertia matrix, $C(q,\dot{q})\dot{q} \in R^n$ is a vector of centripetal and Coriolis generalized forces, $g(q) \in R^n$ is a vector of generalized gravitational forces, $V \in R^n$ is a vector of motor voltages, $K_m, K_b, R, J_m, B_m, r \in R^{n \times n}$ are constant diagonal matrices of torque constant, back emf constant, resistance, inertia, damping and reduction gear ratio of motors, respectively.

3. FUZZY PD + I CONTROLLER

Fuzzy PID controller is implemented as fuzzy PD + I controller as shown in Figure 1. Each fuzzy set consists of a number of MFs to describe the heuristic variables in a mathematical manner. The motor voltage is the output of fuzzy controller while the joint position error and its derivative are its inputs. MFs for the inputs and output of the controller are five fuzzy sets namely NB (Negative Big), NM (Negative Medium), Z (Zero), PM (Positive Medium) and PB (Positive Big).

The following assumptions are given to design the FLC:

- MFs are triangular specified with three points.

- The physical range of inputs are scaled between [−1, 1].

- Axes of the first and the last MF are at −1 and 1, respectively.

- Number of fuzzy sets is an odd integer greater than one.

- MFs are arranged such that the second point of each MF is coincident with the third point of the left one and the first point of the right one.

Then, number and position of second point of MFs are selected as two design parameters.

Position of the MF is specified by spacing parameter where one indicates an even spacing, while any value larger than unity indicates that the MFs are close together in the center of the range and more spaced out at the extremes as shown in Figure 2. This method of designing the MFs is introduced in [16].

The rule-base also is designed based on the ideas presented in [16]. In specifying a rule base, characteristic spacing parameter for each variable and characteristic angle for each input variable are used to construct the rules. The characteristic spacing parameters and the characteristic angle determine how the space is partitioned. The angle determines the slope of a line through the origin on which seed points are placed. The positioning of the seed points is determined by a similar spacing method as was used to determine the centers of the MFs as illustrated in Figure 3 where seed points are blue circles and grid-points are red circles. The lines on the graph delineate the different regions corresponding to different consequents. The parameters for this example are 1 for both input spacing, 0.85 for the output spacing and 40° for the angle. Table 1 shows the derived fuzzy rules.

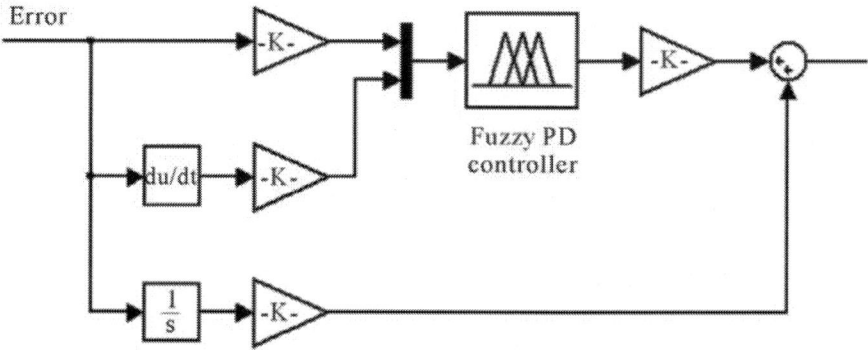

Figure 1. Fuzzy PD + I controller.

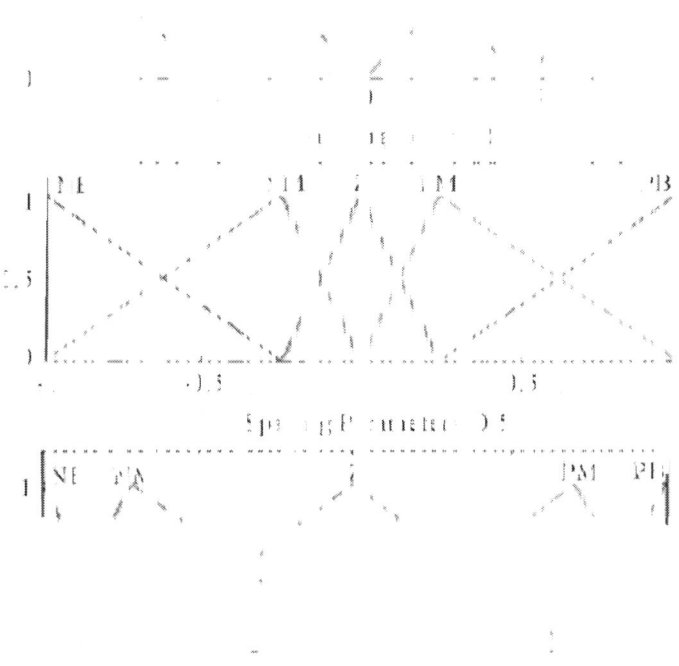

Figure 2. Effect of spacing parameter on MFs.

4. BACTERIAL FORAGING OPTIMIZATION ALGORITHM

The BFO algorithm can be explained by four processes namely, chemotaxis, swarming, reproduction, and elimination-dispersal [6]. Below we briefly describe each of these processes.

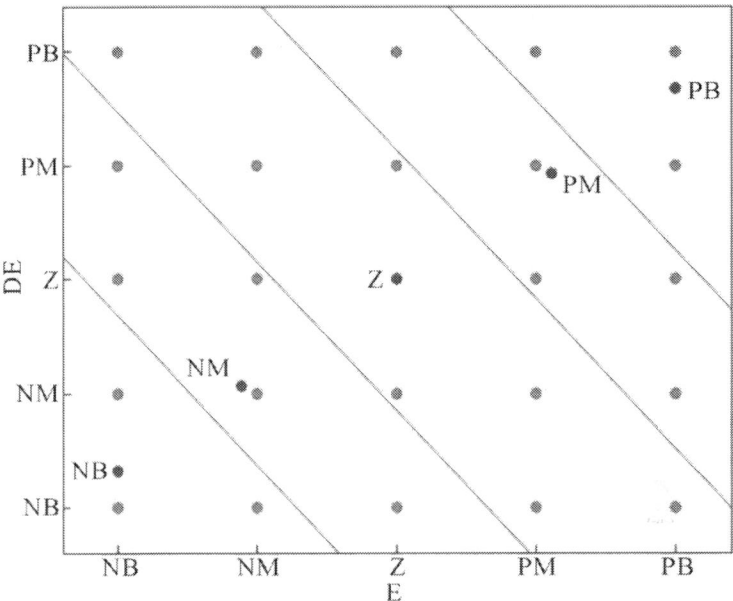

Figure 3. Sample decision plane.

Table 1. Fuzzy rules.

		E				
		NB	NM	Z	PM	PB
	NB	NB	NB	NM	Z	Z
	NM	NB	NM	Z	Z	PM
DE	Z	NM	NM	Z	PM	PM
	PM	NM	Z	Z	PM	PB
	PB	Z	Z	PM	PB	PB

Chemotaxis: This process simulates the swimming and tumbling movements of an E. coli cell by a set of rigid flagella. An E. coli bacterium can move in two different ways. It can swim for a period of time in the same direction or it may tumble, and alternate between these two modes of operation for the entire lifetime. This alternation between the two modes enables the bacterium to move in random directions and search for nutrients. Suppose $\theta^i(j,k,l)$ represents i-th bacterium at j-th chemotactic, k-th reproductive and l-th elimination-dispersal step. C(i) is the run length which is a constant in basic BFO algorithm. In computational chemotaxis, the movement of the bacterium is represented as

$$\theta^i\left(j+1,k,l\right) = \theta^i\left(j,k,l\right) + C(i)\Delta(i)/\sqrt{\Delta^T(i)\Delta(i)} \tag{2}$$

where Δ indicates a vector in the random direction whose elements lie in [−1, 1].

Swarming: It is always desired that when any one of the bacteria reaches the better location, try to attract other bacteria so that they reach the desired place more rapidly. The effect of swarming is to make the bacteria congregate into groups and move as concentric patterns with high bacterial density. Mathematically, swarming can be represented by

$$\begin{aligned}
J_{cc}&\left(\theta, P(j,k,l)\right) \\
&= \sum_{i=1}^{S} J_{cc}\left(\theta, \theta^i\left(j,k,l\right)\right) \\
&= \sum_{i=1}^{S}\left[-d_{attractant}\exp\left(-w_{attractant}\sum_{m=1}^{p}\left(\theta_m - \theta_m^i\right)^2\right)\right] \\
&\quad + \sum_{i=1}^{S}\left[-h_{repellant}\exp\left(-w_{repellant}\sum_{m=1}^{p}\left(\theta_m - \theta_m^i\right)^2\right)\right]
\end{aligned} \tag{3}$$

where $J_{cc}(h,P(j,k,l))$ represents the objective function value to be added to the actual objective function (to be minimized) to present a time varying objective function, S is the total number of bacteria, p is the number of variables to be optimized, which are present in each bacterium and $\theta = \left[\theta_1, \theta_2, \cdots, \theta_p\right]^T$ is a point in the p-dimensional search domain. dattractant, wattractant, hrepellant, wrepellant are different coefficients that should be chosen properly.

Reproduction: The least healthy bacteria eventually die while each of the healthier bacteria (each with the lower cost function) asexually split into two bacteria, which are then placed in the same location. Thus, the population size after reproduction is maintained constant.

Elimination and dispersal: A gradual or sudden changes in the location where a bacterium population lives may occur due to noxious substance, the temperature rises abruptly in the area or some other influence. Events can kill or disperse all the bacteria in a region. This reduces the chances of convergence at local optima location. To simulate this phenomenon in BFO algorithm, some bacteria are chosen, according to apreset probability P_{ed}, to be dispersed and moved to another position within the environment.

The BFO algorithm parameters are denoted as p, S, N_c, N_s, N_{re}, N_{ed}, P_{ed}, where p is dimension of the search space, S is the total number of bacteria in the population, N_c is number of chemotactic steps, N_s is swimming length, N_{re} is number of reproduction steps, N_{ed} is number of elimination-dispersal events, P_{ed} is elimination-dispersal probability. The parameters selected for the proposed BFO algorithm are shown in Table 2.

It is certainly impossible to explore all the potential uses of BFO in this single article, but we briefly point to the some of them. It should seem at least plausible that there are applications of the methods to optimization, optimal control, adaptive estimation and control, and model predictive control.

4.1. Particle Swarm Optimization

PSO is a population-based optimization method inspired by the social behavior of animals such as bird flocking and fish schooling. Like evolutionary algorithms, PSO algorithm conducts search using a population of particles, corresponding to individuals. Each particle has a velocity vector v^i and a position vector x^i to represent a possible solution to the optimization problem. The first positions and velocities of a PSO algorithm are randomly initialized within a population. At the next iteration, position and velocity of each particle are updated by the two values. The first value, pbest, is the personal best position of particle that it has achieved so far. The other, gbest, is obtained by choosing the overall best value from all particles. The new velocity for each particle is updated by the following equation

$$v_n^{(t+1)} = w v_n^{(t)} + c_1 r_1 \left(\text{pbest}_d - x_d^{(t)} \right) + c_2 r_2 \left(\text{gbest}_d - x_d^{(t)} \right)$$

(4)

where w, c_1 and c_2 are called the coefficient of inertia, cognitive and society study, respectively. The r_1 and r_2 is uniformly distributed random numbers in [0, 1].

Changing velocity enables every particle to search around its individual best position and global best position. Based on the updated velocities, each particle changes its position as following

$$x_n^{(t+1)} = x_n^{(t)} + \chi v_n^{(t)}$$

(5)

Table 2. Parameters of BFO algorithm.

S	N_c	N_s	N_{re}	N_{ed}	P_{ed}
10	30	4	5	4	0.25

4.2. Optimization of FLC Using BFO Algorithm

In this paper, spacing parameter for MFs of input/output variables, spacing parameters and angle parameters for rule base and input/output scaling factors of FLCs are determined with BFO algorithm. Figure 4 illustrates the block structure of the FLC optimizing process using BFO algorithm. All parameters of the FLC are updated at every final time (t_f). The method of tuning PID parameters is based upon minimizing the ITAE of joints. If $q_d(k)$ is desired trajectory and $q(k)$ is output trajectory then error $e(k)$ is

$$e(k) = q_d(k) - q(k)$$

(6)

$$ITAE = \sum_{j=1}^{3} \sum_{k=1}^{n} \left| e(kj) \cdot kj \right| \tag{7}$$

where e(kj) is the system error at k-th sampling instant for j-th joint. Tuning process of FLC parameters with PSO is similar to BFO algorithm.

5. SIMULATION RESULTS

The objective of this section is to verify the performances of the Fuzzy-BFO based controller and the Fuzzy-PSO based controller. Results of the tuning methods are tested in terms of ITAE in joint space and ISE in Cartesian space. The desired Cartesian space trajectory is a spiral path. Integral Square Errors ISEX, ISEY and ISEZ are calculated to compare controller performance stated as follows

$$ISEX = \sum_{k=1}^{n} \left(x_d(k) - x(k) \right)^2$$

$$ISEY = \sum_{k=1}^{n} \left(y_d(k) - y(k) \right)^2$$

$$ISEZ = \sum_{k=1}^{n} \left(z_d(k) - z(k) \right)^2 \tag{8}$$

where $\left[x_d(k), y_d(k), z_d(k) \right]$ and $\left[x(k), y(k), z(k) \right]$ are desired and output Cartesian space points at k-th sampling instant. The ability of control system for rejecting disturbances is simulated. For checking the robustness of controller a disturbance torque D is applied as an example in the form of

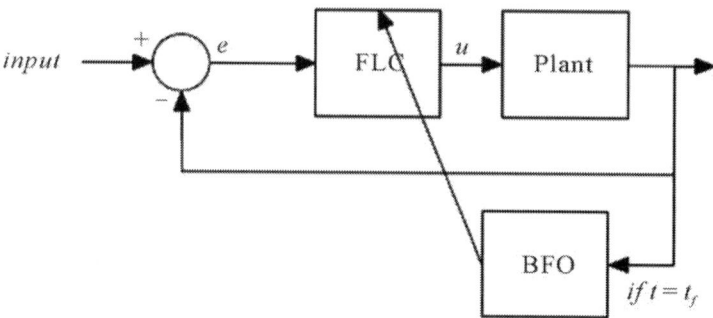

Figure 4. Tuning of FLC parameters by BFO algorithm.

$$D = 29\sin(2t) - 5$$

(9)

Initial angles of all three joints are set to zero, however spiral trajectory is starting from a non-zero value in Figure 6. Fitness function curve of PSO tuning is shown in Figure 5. Performance of the second joint for tracking spiral trajectory and corresponding error are shown in Figures 6 and 7, respectively. We can see a good tracking, where the tracking error of the joint is very small. Tables 3 and 4 give the values of the two cost functions, ISE and ITAE, with and without disturbance for the controllers. As can be seen, the Fuzzy-BFO based controller is better than the Fuzzy-PSO based controller. Considering the results confirm a powerful ability of rejecting disturbances. Figures 8 and 9 show desired and tracked trajectory for tuned controllers. With comparing performances in simulations, it can be concluded that the BFO algorithm is superior to the PSO algorithm in term of accuracy of response.

6. CONCLUSION

In this paper, we have presented a comparison study of using BFO and PSO algorithms for a design of a fuzzy PID controller to tracking control. Performances of controllers in the cases of with and without disturbances are

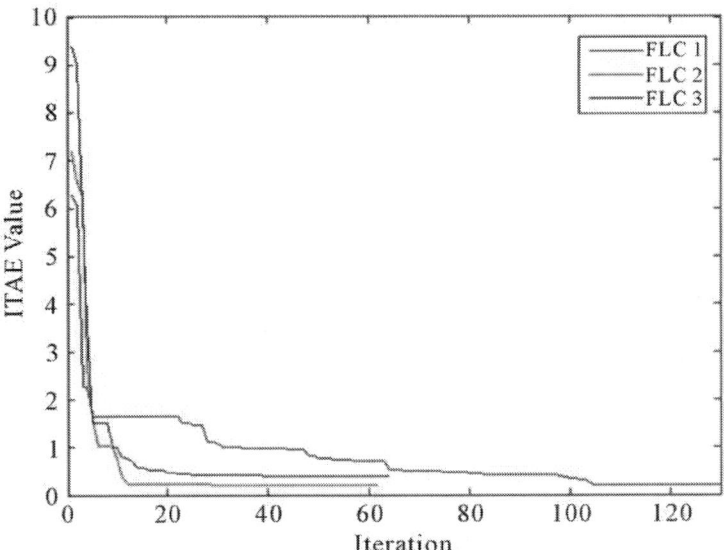

Figure 5. Fitness function in PSO algorithm.

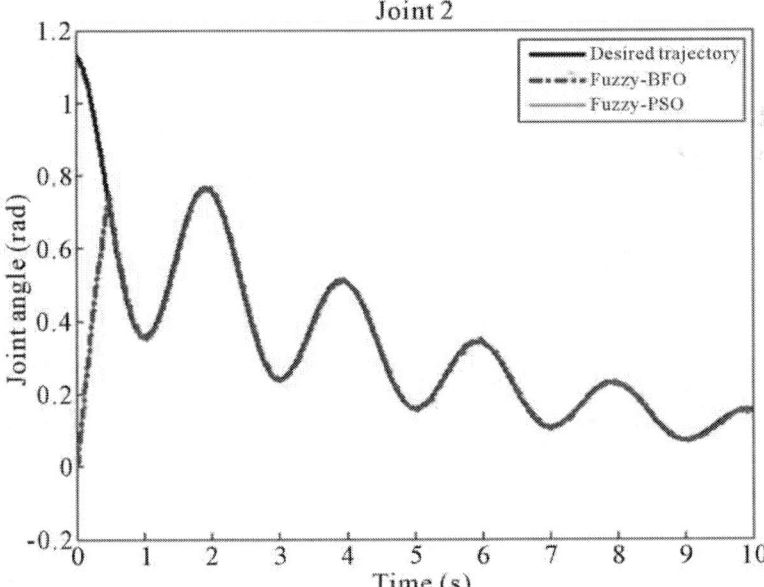

Figure 6. Performance of second joint angle.

Table 3. Fitness function (ISE) in Cartesian space.

	Without Disturbance			With Disturbance		
	ISEX	ISEY	ISEZ	ISEX	ISEY	ISEZ
BFO	0.1684	0.0397	0.0147	0.1689	0.0398	0.0149
PSO	0.1807	0.0452	0.0934	0.1809	0.0453	0.0947

Table 4. Fitness function (ITAE) in joint space.

	Without Disturbance	With Disturbance
BFO	0.6255	0.6378
PSO	0.9342	0.9379

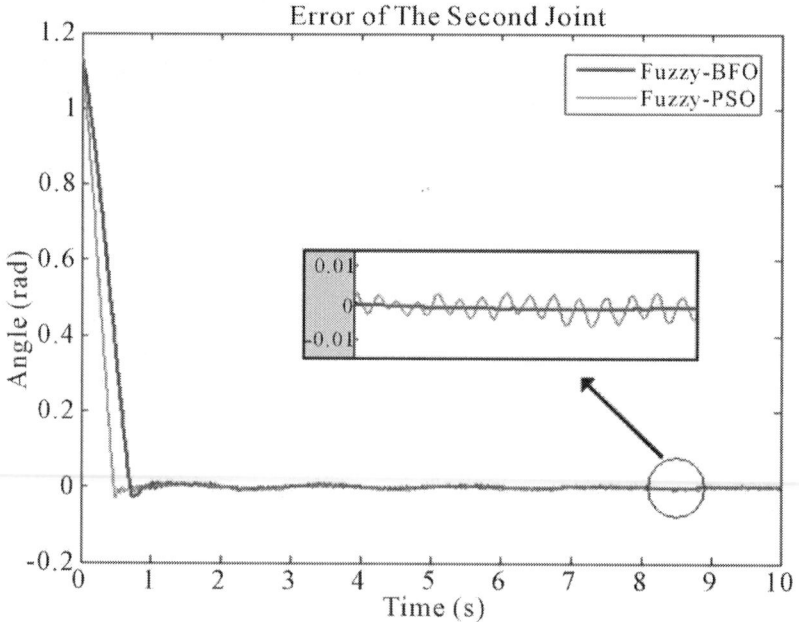

Figure 7. Joint space error of second joint.

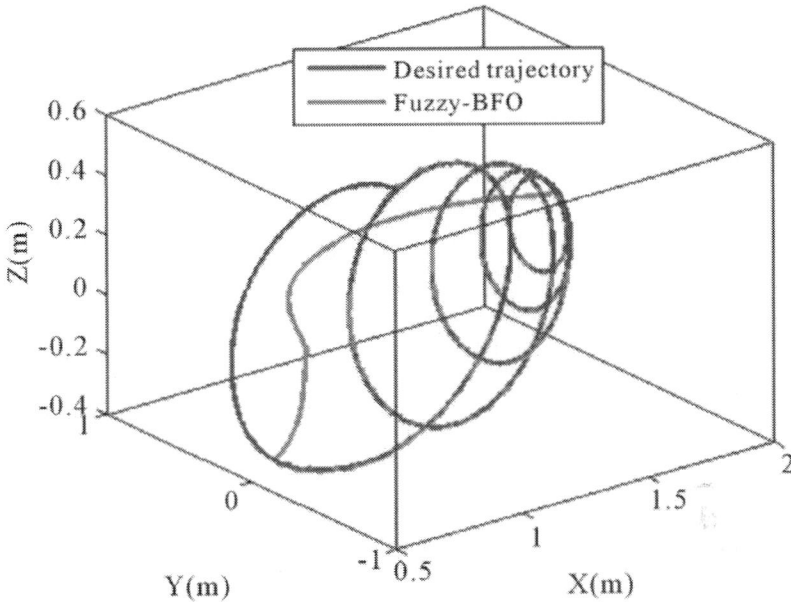

Figure 8. Performance of FLC using BFO algorithm.

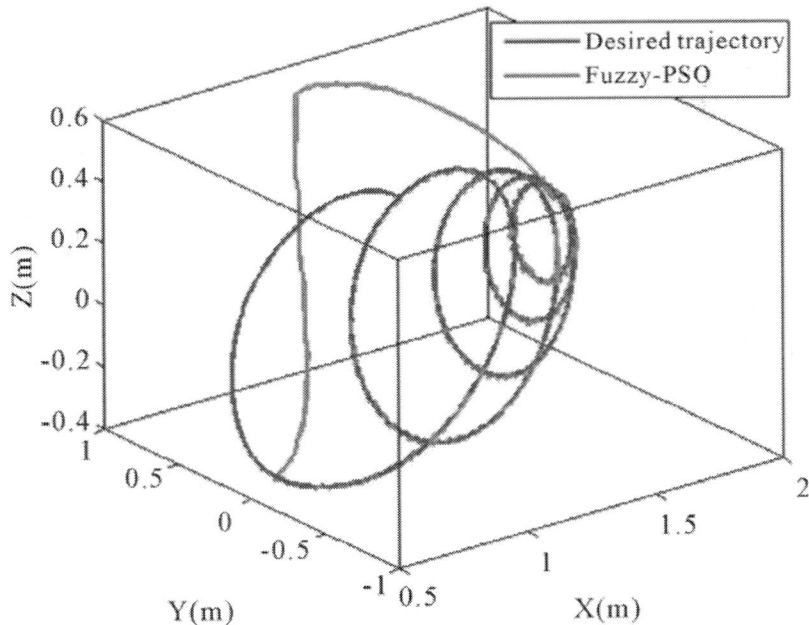

Figure 9. Performance of FLC using PSO algorithm.

compared for the above approaches in joint space, as well as in Cartesian space. The simulation results show that BFO algorithm is superior to PSO algorithm in term of accuracy of response. An improvement of this work can be made by designing an online adaptive controller based on BFO algorithm.

REFERENCES

1. K. H. Ang, G. Chong and Y. Li, "PID Control System Analysis, Design, and Technology," IEEE Transactions on Control Systems Technology, Singapore, Vol. 13, No. 4, 2005, pp. 559-576.

2. T. H. Kim, I. Maruta and T. Sugie, "Robust PID Controller Tuning Based on the Constrained Particle Swarm Optimization," Automatica, Vol. 44, No. 4, 2008, pp. 1104- 1110.

3. L. X. Wang, "A Course in Fuzzy Systems and Control," Prentice Hall, New York, 1996.

4. M. M. Fateh, "Robust Fuzzy Control of Electrical Manipulators," Journal of Intelligent and Robotic Systems, Vol. 60, No. 3-4, 2010, pp. 415-434.

5. K. M. Passino, "Biomimicry of Bacterial Foraging for Distributed Optimization and Control," IEEE Control Systems Magazine, Columbus, Vol. 22, No. 3, 2002, pp. 52-67.

6. J. Kennedy and R. Eberhart, "Particle Swarm Optimization," IEEE International Conference on Neural Networks, Vol. 4, Perth, 27 November-1 December 1995, pp. 1942-1948.

7. M. Dorigo and T. Stutzle, "Ant Colony Optimization," MIT Press, Cambridge, 2004.

8. M. Tripathy and S. Mishra, "Bacteria Foraging-Based to Optimize Both Real Power Loss and Voltage Stability Limit," IEEE Transactions on Power Systems, Vol. 22, No. 1, 2007, pp. 240-248.

9. M. A. Munoz, J. A. Lopez and E. Caicedo, "Bacteria Swarm Foraging Optimization for Dynamical Resource Allocation in A Multizone Temperature Experimentation Platform," Analysis and Design of

Intelligent Systems using Soft Computing Techniques, Advances in Intelligent and Soft Computing, Vol. 41, 2007, pp. 427-435.

10. H. Shen, Y. Zhu, X. Zhou, H. Guo and C. Chang, "Bacterial Foraging Optimization Algorithm with Particle Swarm Optimization Strategy for Global Numerical Optimization," Proceedings of the First ACM/SIGEVO Summit on Genetic and Evolutionary Computation, Shanghai, June 2009, pp. 497-504.

11. D. H. Kim, A. Abraham and J. H. Cho, "A Hybrid Genetic Algorithm and Bacterial Foraging Approach for Global Optimization," Information Sciences, Vol. 177, No. 18, 2007, pp. 3918-3937.

12. T. J. Su, G. Y. Chen, J. C. Cheng and C. J. Yu, "Fuzzy PID Controller Design Using Synchronous Bacterial Foraging Optimization," 3rd International Conference on Information Sciences and Interaction Sciences, Kaohsiung, June 2010, pp. 639-642.

13. A. Biswas, S. Dasgupta, S. Das and A. Abraham, "Synergy of PSO and Bacterial Foraging Optimization: A Comparative Study on Numerical Benchmarks," Innovations in Hybrid Intelligent Systems, Advances in Intelligent and Soft Computing, Vol. 44, 2007, pp. 255-263.

14. M. Clerc and J. Kennedy, "The Particle Swarm-Explosion, Stability and Convergence in A Multidimensional Complex Space," IEEE Transactions on Evolutionary Computation, Vol. 6, No. 1, 2002, pp. 58-73.

15. M. M. Fateh, "Proper Uncertainty Bound Parameter to Robust Control of Electrical Manipulators Using Nominal Model," Nonlinear Dynamics, Vol. 61, No. 4, 2010, pp. 655-666.

16. Y. J. Park, H. S. Cho and D. H. Cha, "Genetic AlgorithmBased Optimization of Fuzzy Logic Controller Using Characteristic Parameters," IEEE International Conference on Evolutionary Computation, Taejon, Vol. 2, 1995, pp. 831-836.

CHAPTER 3

Optimal Usage of Robot Manipulators

Behnam Kamrani[1], Viktor Berbyuk[2], Daniel Wäppling[3], Xiaolong Feng[4] and Hans Andersson[4]

[1]MSC.Software Sweden AB, SE-42 677, Gothenburg
[2]Chalmers University of Technology, SE-412 96, Gothenburg
[4]ABB Corporate Research, SE-72178, Västerås
[3]ABB Robotics, SE-78 168, Västerås Sweden

1. INTRODUCTION

Robot-based automation has gained increasing deployment in industry. Typical application examples of industrial robots are material handling, machine tending, arc welding, spot welding, cutting, painting, and gluing. A robot task normally consists of a sequence of the robot tool center point (TCP) movements. The time duration during which the sequence of the TCP movements is completed is referred to as cycle time. Minimizing cycle time implies increasing the productivity, improving machine utilization, and thus making automation affordable in applications for which throughput and cost effectiveness is of major concern. Considering the high number of task runs within a specific time span, for instance one year, the importance of reducing cycle time in a small amount such as a few percent will be more understandable.

Robot manipulators can be expected to achieve a variety of optimum objectives. While the cycle time optimization is among the areas which have probably received the most attention so far, the other application aspects such as energy efficiency, lifetime of the manipulator, and even the environment aspect have also gained increasing focus. Also, in recent era virtual product development technology has been inevitably and enormously deployed toward achieving optimal solutions. For example, off-line programming of robotic work- cells has become a valuable means for work-cell designers to investigate the manipulator's workspace to achieve optimality in cycle time, energy consumption and manipulator lifetime.

This chapter is devoted to introduce new approaches for optimal usage of robots. Section 2 is dedicated to the approaches resulted from translational and rotational repositioning of a robot path in its workspace based on response surface method to achieve optimal cycle time. Section 3 covers another proposed approach that uses a multi-objective optimization methodology, in which the position of task and the settings of drive-train components of a robot manipulator are optimized simultaneously to understand the trade-off among cycle time, lifetime of critical drive-train components, and energy efficiency. In both section 2 and 3, results of different case studies comprising several industrial robots performing different tasks are presented to evaluate the developed methodologies and algorithms. The chapter is concluded with evaluation of the current results and an outlook on future research topics on optimal usage of robot manipulators.

2. TIME-OPTIMAL ROBOT PLACEMENT USING RESPONSE SURFACE METHOD

This section is concerned with a new approach for optimal placement of a prescribed task in the workspace of a robotic manipulator. The approach is resulted by applying response surface method on concept of path translation and path rotation. The methodology is verified by optimizing the position of several kinds of industrial robots and paths in four showcases to attain minimum cycle time.

2.1 Research background

It is of general interest to perform the path motion as fast as possible. Minimizing motion time can significantly shorten cycle time, increase the productivity, improve machine utilization, and thus make automation affordable in applications for which throughput and cost effectiveness is of major concern.

In industrial application, a robotic manipulator performs a repetitive sequence of movements. A robot task is usually defined by a robot program, that is, a robot pathconsisting of a set of robot positions (either joint positions or tool center point positions) and corresponding set of motion definitions between each two adjacent robot positions. *Path translation* and *path rotation* terms are repeatedly used in this section to describe the methodology. Path translation implies certain translation of the path in x, y, z directions of an arbitrary coordinate system relative to the robot while all path points are fixed with respect to each other. Path rotation implies certain rotation of the path with θ, φ, ψ angles of an arbitrary coordinate system relative to the robot while all path points are fixed with respect to each other. Note that since path translation and path rotation are relative concepts, they may be achieved either by relocating the path or the robot.

In the past years, much research has been devoted to the optimization problem of designing robotic work cells. Several approaches have been used in order to define the optimal relative robot and task position. A *manipulability measure* was proposed (Yoshikawa, 1985) and a modification to Yoshikawa's manipulability measure was proposed (Tsai, 1986) which also accounted for proximity to joint limits. (Nelson & Donath, 1990) developed a gradient function of manipulability in Cartesian space based on explicit determination of manipulability function and the gradient of the manipulability function in joint space.

Then they used a modified method of the steepest descent optimization procedure (Luenberger, 1969) as the basis for an algorithm that automatically locates an assembly task away from singularities within manipulator's workspace.

In aforementioned works, mainly the effects of robot kinematics have been considered.Once a robot became employed in more complex tasks requiring improved performance, e .g., higher speed and accuracy of trajectory tracking, the need for taking into account robot dynamics becomes more essential (Tsai, 1999).

A study of time-optimal positioning of a prescribed task in the workspace of a 2R planar manipulator has been investigated (Fardanesh & Rastegar, 1988). (Barral et al., 1999) applied the simulated annealing optimization method to two different problems: robot placement and point-ordering optimization, in the context of welding tasks with only one restrictiveworking hypothesis for the type of the robot. Furthermore, a state of the art of different methodologies has been presented by them.

In the current study, the dynamic effect of the robot is considered by utilizing a computer model which simulates the behavior and response of the robot, that is, the dynamic models of the robots embedded in ABB's IRC5 controller. The IRC5 robot controller uses powerful, configurable software and has a unique dynamic model-based control system which provides self-optimizing motion (Vukobratovic, 2002).

To the best knowledge of the authors, there are no studies that directly use the response surface method to solve optimization problem of optimal robot placement considering a general robot and task. In this section, a new approach for optimal placement of a prescribed task in the workspace of a robot is presented. The approach is resulted by path translation and path rotation in conjunction with response surface method.

2.2 Problem statement and implementation environment

The problem investigated is to determine the relative robot and task position with the objective of time optimality. Since in this study a relative position is to be pursued, either the robot, the path, or both the robot and path may be relocated to achieve the goal. In such a problem, the robot is given and specified without any limitation imposed on the robot type, meaning that any kind of robot can be considered. The path or task, the same as the robot, is given and specified; however, the path is also general and any kind of path can be considered. The optimization objective is to define the optimal relative position between a robotic manipulator and a path. The optimal location of the task is a location which yields a minimum cycle time for the task to be performed by the robot.

To simulate the dynamic behavior of the robot, RobotStudio is employed, that is a software product from ABB that enables offline programming and simulation of robot systems using a standard Windows PC. The entire robot, robot tool, targets, path, and coordinate systems can be defined and specified in RobotStudio. The simulation of a robot system in RobotStudio employs the ABB Virtual Controller, the real robot program, and the configuration file that are identical to those used on the factory floor. Therefore the simulation predicts the true performance of the robot.

In conjunction with RobotStudio, Matlab and Visual Basic Application (VBA) are utilized to develop a tool for proving the designated methodology. These programming environments interact and exchange data with each other simultaneously. While the main dataflow runs in VBA, Matlab stands for numerical computation, optimization calculation, and post processing. RobotStudio is employed for determining the *path admissibility* boundaries and calculating the cycle times. Figure 1 illustrates the schematic of dataflow in the three computational environments.

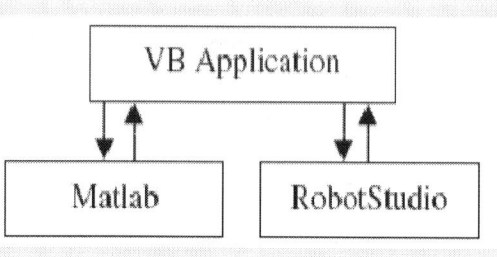

Figure. 1. Dataflow in the three computational tools

2.3 Methodology of time-optimal robot placement

Basically, the path position relative to the robot can be modified by translating and/or rotating the path relative to the robot. Based on this idea, translation and rotation approaches are examined to determine the optimal path position. The algorithms of both approaches are considerably analogous. The approaches are based on the response surface method and consist of following steps. First is to pursue the *admissibility boundary*, that is, the boundary of the area in which a specific task can be performed with the same robot configuration as defined in the path instruction. This boundary is obviously a subset of the general robot operability space that is specified by the robot manufacturer. The computational time of this step is very short and may take only few seconds. Then experiments are performed on different locations of admissibility boundary to calculate the cycle time as a function of path location. Next, optimum path location is determined by using constrained optimization technique implemented in Matlab. Finally, the sensitivity analysis is carried out to increase the accuracy of optimum location.

Response surface method (Box et al., 1978; Khuri & Cornell, 1987; Myers & Montgomery, 1995) is, in fact, a collection of mathematical and statistical techniques that are useful for the modeling and analysis of problems in which a response of interest is influenced by several decision variables and the objective is to optimize the response. Conventional optimization methods are often cumbersome since they demand rather complicated calculations, elaborate skills, and notable simulation time. In contrast, the response surface method requires a limited number of simulations, has no convergence issue, and is easy to use.

In the current robotic problem, the decision variables consist of x, y, and z of the reference coordinates of a prescribed path relative to a given robot base and the response of interest to be minimized is the task cycle time. A so-called full factorial design is considered by 27 experiment points on the path admissibility boundaries in three-dimensional space with original path location in center. Figure 2 graphically depicts the original path location in the center of the cube and the possible directions for finding the admissibility boundary.

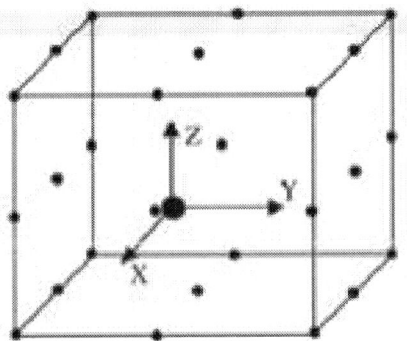

Figure. 2. Direction of experiments relative to the original location of path

Three-dimensional bisection algorithm is employed to determine the path admissibility region. The algorithm is based on the same principle as the bisection algorithm for locating the root of a three-variable polynomial. Bisection algorithm for finding the admissibility boundary states that each translation should be equal to half of the last translation and translation direction is the same as the last translation if all targets in the path are admissible; otherwise, it is reverse. Herein, targets on the path are considered admissible if the robot manipulator can reach them with the predefined configurations. Note that in this step the robot motion between targets is not checked.

Since the target admissibility check is only limited to the targets and the motion between the targets are not simulated, it has a low computational cost. Additionally, according to practical experiments, if all targets are admissible, there is a high probability that the whole path would also be admissible. However, checking the target admissibility does not guarantee that the whole path is admissible as the joint limits must allow the manipulator to track the path between the targets as well. In fact, for investigating the path admissibility, it is necessary to simulate the whole task in RobotStudio to ascertain that the robot can manage the whole task, i.e., targets and the path between targets.

To clarify the method, an example is presented here. Let's assume an initial translation by

1.0 m in positive direction of x axis of reference coordinate system is considered. If all targets after translation are admissible, then the next translation would be 0.5 m and in the same (+x) direction; otherwise in opposite (−x) direction. In any case, the admissibility of targets in the new location is checked and depending on the result, the direction for the next translation is decided. The amount of new translation would be then 0.25 m. This process continues until a location in which all targets are admissible is found such that the last translation is smaller than a certain value, that is, the considered tolerance for finding the boundary, e.g., 1 mm.

After finding the target admissibility boundary in one direction within the decided tolerance, a whole task simulation is run to measure the cycle time. Besides measuring the cycle time, it is also controlled if the robot can perform the whole path, i.e., investigating the *path admissibility* in addition to targets admissibility. If the path is not admissible in that location, a new admissible location within a relaxed tolerance can be sought and examined. The same procedure is repeated in different directions, e.g. 27 directions in full-factorial method, and by that, a matrix of boundary coordinates and vector of the corresponding cycle times are casted.

A quadratic approximation function provides proper result in most of response surface method problems (Myers & Montgomery, 1995), that is:

$$f(x,y,z) = b_0 + b_1x + b_2y + b_3z + \dots \quad \text{(linear terms)}$$
$$b_4xy + b_5yz + b_6xz + \dots \quad \text{(interaction terms)}$$
$$b_7x^2 + b_8y^2 + b_9z^2 \quad \text{(quadratic terms)}$$

By applying the following mapping

$$x = x_1 ; \quad y = x_2 ; \quad z = x_3$$
$$xy = x_4 ; \quad xz = x_5 ; \quad yz = x_6$$
$$x^2 = x_7 ; \quad y^2 = x_8 ; \quad z^2 = x_9$$

Eq. 1 can be expressed in linear form and by matrix notation as:

$$Y = XB + e$$

where Y is the vector of cycle times, X is the design matrix of boundaries, B is the vector of unknown model coefficients of $\{b0, b1, b2, \dots, b9\}$, and e is the vector of errors. Finally, B can be estimated using the least squares method, minimizing of $L = e^Te$, as:

$$B = (X^TX)^{-1} X^TY$$

In the next step of the methodology, when the expression of cycle time as a function of a reference coordinate (x, y, z) is given, the minimum of the cycle times subject to the determined boundaries is to be found. The *fmincon* function in Matlab optimization toolbox is used to obtain the minimum of a constrained nonlinear function. Note that, since the cycle time function is a prediction of the cycle time based on the limited experiments data, the obtained value (for the minimum of cycle time) does not necessarily provide the global minimum cycle time of the task. Moreover, it is not certain yet that the task in optimum location is kinematically admissible. Due to these reasons, the minimum of the cycle time function can merely be considered as an *optimum candidate.*

Hence, the optimum candidate must be evaluated by performing a confirmatory task simulation in order to, first investigate whether the location is admissible and second, calculate the actual cycle time. If the location is not admissible, the closest location in the direction of the translation vector is pursued such that all targets are admissible. This new location is considered as a new optimum candidate and replaced the old one. This procedure may be called sequential backward translation.

Due to the probability of inadmissible location and as a work around, the algorithm, by default, seeks and introduces several optimum candidates by setting different search areas in *fmincon* function. All candidate locations are examined and cycle times are measured. If any location is inadmissible, that location is removed from the list of optimum candidate. After examining all the candidates, the minimum value is selected as the final optimum. If none of the optimum candidates is admissible, the shortest cycle time of experiments is selected as optimum. In fact, and in any case, it is always reasonable to inspect if the optimum cycle time is shorter than all the experiment cycle times, and if not, the shortest cycle time is chosen as the local optimum.

As the last step of the methodology the sensitivity analysis of the obtained optimal solution with respect to small variations in x, y, z coordinates can be interesting to study. This analysis can particularly be useful when other constraints, for example space inadequacy, delimit the design of robotic cell. Another important benefit of this analysis is that it usually increases the accuracy of optimum location, meaning that it can lead to finding a precise local optimum location.

The sensitivity analysis procedure is generally analogous to the main analysis. However, herein, the experiments are conducted in a small region around the optimum location. Also, note that since it is likely that the optimum point, found in the previous step, is located on (or close to) the boundary, defining a cube around a point located on the boundary places some cube sides outside the boundary. For instance, when the shortest cycle time of the experiments is selected as the local optimum, the optimum location is already on the admissibility boundary. In such cases, as a work around, the nearest admissible location in the corresponding direction is considered instead.

Note that the sensitivity analysis may be repeated several times in order to further improve the results. Figure 3 provides an overview of the optimization algorithm.

As was mentioned earlier, the path position relative to the robot can be modified by translating as well as rotating the path. In path translation, the optimal position can be achieved without any change in path orientation. However, in path rotation, the optimal path orientation is to be sought. In other words, in path rotation approach the aim is to obtain the optimum cycle time by rotating the path around the x, y, and z axes of a local frame. The local frame is originally defined parallel to the axes of the global reference frame on an arbitrary point. The origin of the local reference frame is called the rotation center. Three sequential rotation angles are used to rotate the path around the selected rotation center. To calculate new coordinates and orientations of an arbitrary target after a path rotation, a target of T on the path is considered in global reference frame of X–Y–Z which is demonstrated in Fig. 4. The target T is rotated in local frame by a rotation vector of θ, φ, ψ which yields the target T'.

If the targets in the path are not admissible after rotating by a certain rotation vector, the boundary of a possible rotation in the corresponding direction is to be obtained based on the bisection algorithm. The matrices of experiments and cycle time response are built in the same way as described in the path translation section and the cycle time expression as a function of rotation angles of θ, φ, ψ is calculated. The optimum rotation angles are obtained using Matlab *fmincon* function. Finally, sensitivity analyses may be performed. A procedure akin to path translation is used to investigate the effect of path rotation on the cycle time.

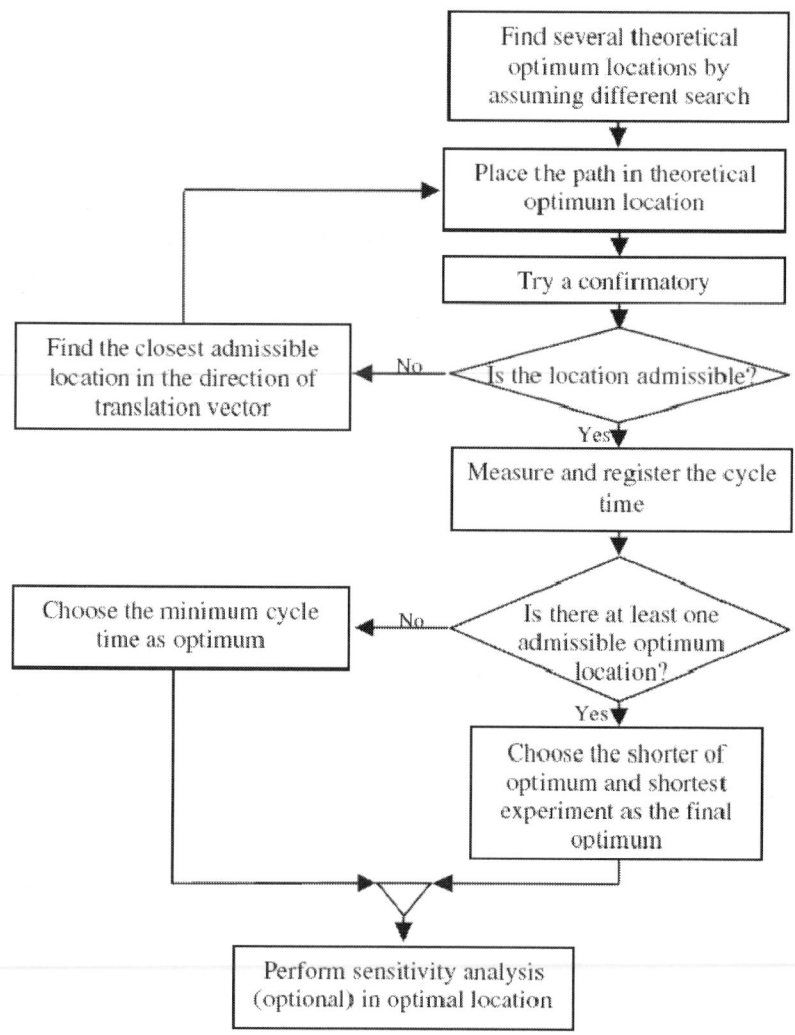

Figure. 3. Flowchart diagram of the optimization algorithm

Although the algorithm of path rotation is akin to path translation, two noticeable differences exist. Although the algorithm of path rotation is akin to path translation, two noticeable differences exist. First, in the rotation approach, the order of rotations must be observed. It can be shown that interchanging orders of rotation drastically influences the resulting orientation. Thus, the order of rotation angles must be adhered to strictly (Haug, 1992). Consequently, in the path rotation approach, the optimal rotation determined by sensitivity analysis cannot be added to the optimal rotation obtained by the main analysis, whereas in the translation approach, they can be summed up to achieve the resultant translation vector. Another difference is that, in the rotation approach, the results logically depend on the selection of the rotation center location, while there is no such dependency in the path translation approach. More details concerning path rotation approach can be found in (Kamrani et al., 2009).

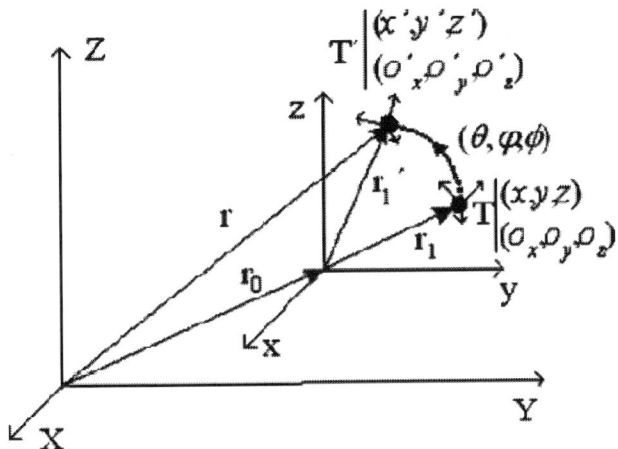

Figure. 4. Rotation of an arbitrary target T in the global reference frame

2.4 Results on time-optimal robot placement

To evaluate the methodology, four case studies comprised of several industrial robots performing different tasks are proved. The goal is to optimize the cycle time by changing the path position. A coordinate system with its origin located at the base of the robot, x-axis pointing radially out from the base, z-axis pointing vertically upwards, is used for all the cases below.

2.4.1 Path Translation

In this section, obtained by path translation approach are presented.

2.4.1.1 Case 1

The first test is carried out using the ABB robot IRB6600-225-175 performing a spot welding task composed of 54 targets with fixed positions and orientations regularly distributed around a rectangular placed on a plane parallel to the x-y plane (parallel to horizon). A view of the robot and the path in its original location is depicted in the Fig. 5. The optimal location of the task in a boundary of $(\pm 0.5\ m,\ \pm 0.8\ m,\ \pm 0.5\ m)$ is calculated using the path translation approach to be $(\Delta x,\ \Delta y,\ \Delta z) = (0\ m,\ 0.8\ m,\ 0\ m)$. The cycle time of this path is reduced from originally 37.7 seconds to 35.7 seconds which implies a gain of 5.3 percent cycle time reduction. Fig. 6 demonstrates the robot and path in the optimal location determined by translation approach.

2.4.1.2 Case 2

The second case is conducted with the same ABB IRB6600-225-175 robot. The path is composed of 18 targets and has a closed loop shape. The path is shown in the Fig. 7 and as can be seen, the targets are not in one plane. The optimal location of the task in a boundary of $(\pm 1.0\ m,\ \pm 1.0\ m,\ \pm 1.0\ m)$ is calculated using the path translation approach to be as $(\Delta x,\ \Delta y,\ \Delta z) = (-0.104\ m,\ -0.993\ m,\ 0.458\ m)$. The cycle time of this path is reduced from originally 6.1 seconds to 5.6 seconds which indicates 8.3 percent cycle time reduction.

2.4.1.3 Case 3

In the third case study, an ABB robot of type IRB4400L10 is considered performing a typical machine tending motion cycle among three targets which are located in a plane parallel to the horizon. The robot and the path are depicted in the Fig. 8. The path instruction states to start from the first target and reach the third target and then return to the starting target. A restriction for this case is that the task cannot be relocated in the y-direction relative to the robot. The optimal location of the task in a boundary of $(\pm 1.0\ m,\ 0\ m,\ \pm 1.0\ m)$ is calculated using the path translation approach to be as $(\Delta x,\ \Delta y,\ \Delta z) = (0.797\ m,\ 0\ m,\ -0.797\ m)$. The cycle time of this path is reduced from originally 2.8 seconds to 2.6 seconds which evidences 7.8 percent cycle time reduction.

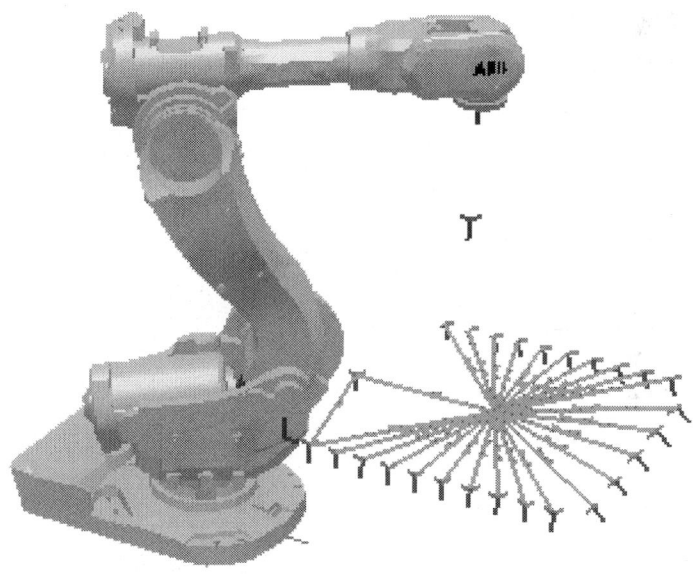

Figure. 5. IRB6600 ABB robot with a spot welding path of case 1 in its original location

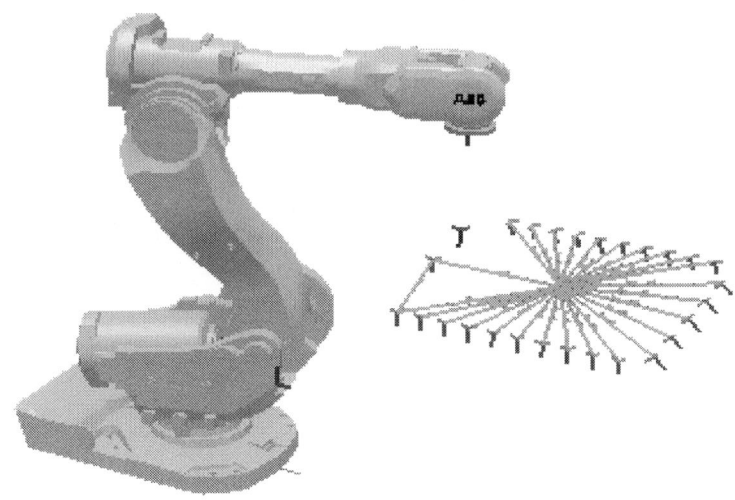

Figure. 6. IRB6600 ABB robot with a spot welding path of case 1 in optimal location found by translation approach

2.4.1.4 Case 4

The forth case is carried out using an ABB robot of IRB640 type. In contrast to the previous robots which have 6 joints, IRB640 has merely 4 joints. The path is shown in the Fig. 9 and comprises four points which are located in a plane parallel to the horizon. The motion instruction requests the robot to start from first point and reach to the forth point and then return to the first point again. The optimal location of the task in a boundary of $(\pm 0.5\ m,\ \pm 0.8\ m,\ \pm 0.5\ m)$ is calculated using the path translation approach to be as $(\Delta x,\ \Delta y,\ \Delta z) = (0\ m,\ 0.8\ m,\ 0\ m)$. The cycle time of this path is reduced from originally 3.7 seconds to 3.5 seconds which gives 5.2 percent cycle time reduction.

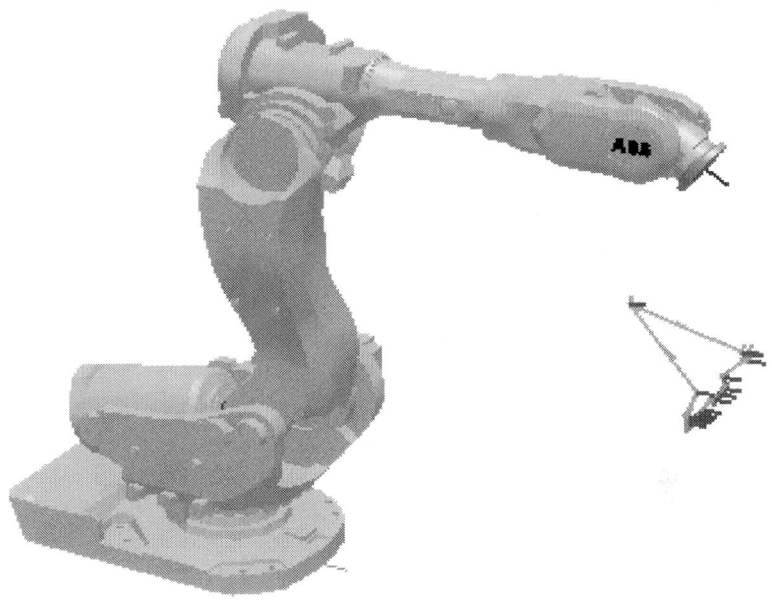

Figure. 7. IRB6600 ABB robot with the path of case 2 in its original location

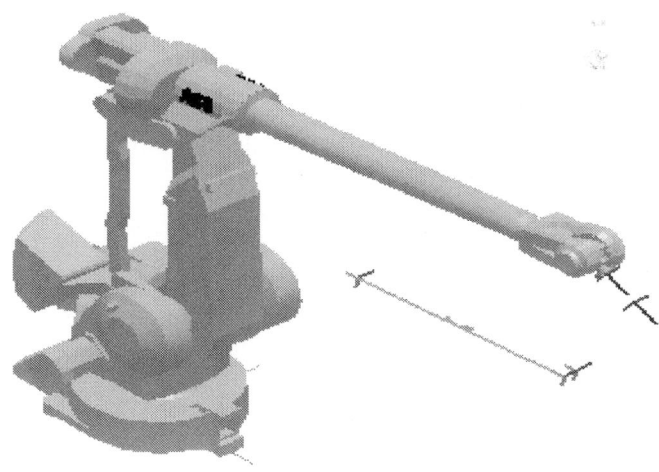

Figure. 8. IRB4400L10 ABB robot with the path of case 3 in its original location

2.4.2 Path Rotation

In this section, results of path rotation approach are presented for four case studies. Herein the same robots and tasks investigated in path translation approach are studied so that comparison between the two approaches will be possible.

2.4.2.1 Case 1

The first case is carried out using the same robot and path presented in section 2.4.1.1. The central target point was selected as the rotation center. The optimal location of the task in a boundary of $(\pm 45°, \pm 45°, \pm 30°)$ is calculated using the path rotation approach to be as $(\Delta\theta, \Delta\varphi, \Delta\psi) = (45°, 0°, 0°)$. The path in the optimal location determined by rotation approach is shown in Fig. 10. The task cycle time was reduced from originally 37.7 seconds to 35.7 seconds which implies an improvement of 5.3 percent compared to the original path location.

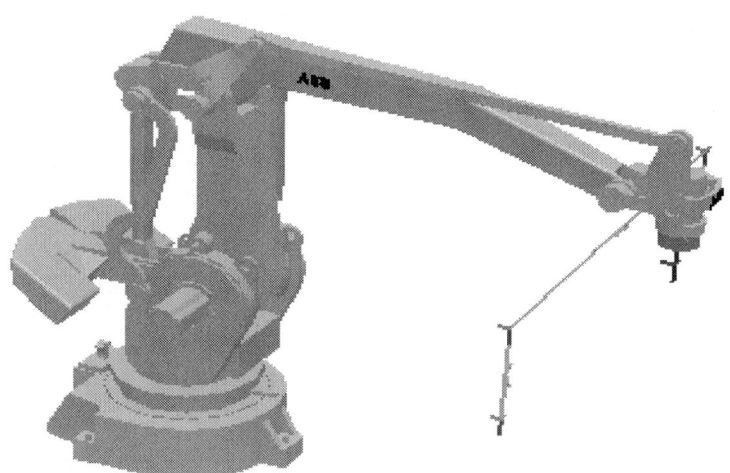

Figure. 9. IRB640 ABB robot with the path of case 4 in its original location

2.4.2.2 Case 2

The second case study is conducted with the same robot and path presented in 2.4.1.2. An arbitrary point close to the trajectory was selected as the rotation center. The optimal location of the task in a boundary of $(\pm 45°, \pm 45°, \pm 30°)$ is calculated using the path rotation approach to be as $(\Delta\theta, \Delta\varphi, \Delta\psi) = (45°, 0°, 0°)$. The cycle time of this path is reduced from originally 6.0 seconds to 5.5 seconds which indicates 8.3 percent cycle time reduction.

2.4.2.3 Case 3

In the third example the same robot and path presented in section 2.4.1.3 are studied. The middle point of the long side was selected as the rotation center. To fulfill the restrictions outlined in section 2.4.1.3, only rotation around y-axis is allowed. The optimal location of the task in a boundary of $(0°, \pm 90°, 0°)$ is calculated using the path rotation approach to be as $(\Delta\theta, \Delta\varphi, \Delta\psi) = (0°, -60°, 0°)$. Here the sensitivity analysis was also performed. The cycle time of this path is reduced from originally 2.8 seconds to 2.2 seconds which evidences 21 percent cycle time reduction.

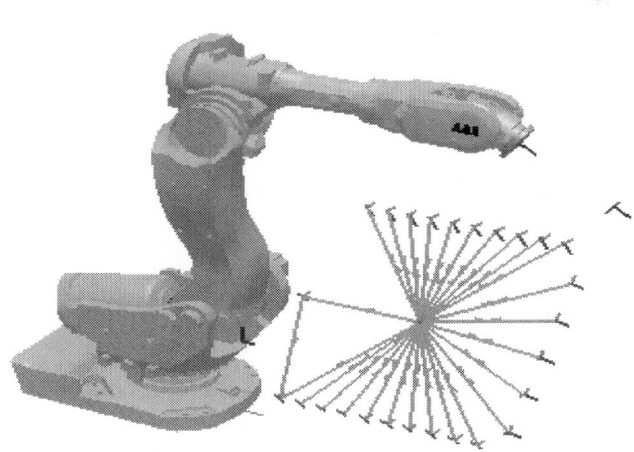

Figure. 10. IRB6600 ABB robot with a spot welding path of case 1 in optimal location found by rotation approach

2.4.2.4 Case 4

The forth case study is carried out with the same robot presented in 2.4.1.4. The point in the middle of a line which connects the first and forth targets was chosen as the rotation center. Due to the fact that the robot has 4 degrees of freedom, only rotation around the z-axis is allowed. The optimal location of the task in a boundary of $(0°, 0°, \pm 45°)$ is calculated using the path rotation approach to be as $(\Delta\theta, \Delta\varphi, \Delta\psi) = (0°, 0°, 16°)$. In this case the sensitivity analysis was also performed. The cycle time of this path is reduced from originally 3.7 seconds to 3.6 seconds which gives 3.5 percent cycle time reduction.

2.4.3 Summary of the Results of Section 2

The cycle time reduction percentages that are achieved by translation and rotation approaches compared to longest and original cycle time are demonstrated in Fig. 11. The longest cycle time which corresponds to worst performance location is recognized as an   existing admissible location that has the longest cycle time, i.e., the longest cycle time among experiments. As can be perceived, a cycle time reduction in range of 8.7 – 37.2 percent is achieved as compared to the location with the worst performance.

Results are also compared with the cycle time corresponding to original path location. This   comparison is of interest as the tasks were programmed by experienced engineers and had been originally placed in proper position. Therefore this comparison can highlight the efficiency and value of the algorithm. The results demonstrate that cycle time is reduced by 3.5 - 21.1 percent compared with the original cycle time.

Fig. 11 indicates that both translation and rotation approaches are capable to noticeably reduce the cycle time of a robot manipulator.

A relatively lower gain in cycle time reduction in case four is related to a robot with four joints. This robot has fewer joint than the other tested robots with six joints. Generally, the fewer number of joints in a robot manipulator, the fewer degrees of freedom the robot has. The small variation of the cycle time in the whole admissibility area can imply that this robot has a more homogeneous dynamic behavior. Path geometry may also contribute to this phenomenon.

Also note that cycle time may be further reduced by performing more experiments. Although doing more experiments implies an increase in simulation time, this cost can reasonably be neglected by noticing the amount of time saving, for instance 20 percent in one year. In other word, the increase in productivity in the long run can justify the initial high computational burden that may be present, noting that this is a onetime effort before the assembly line is set up.

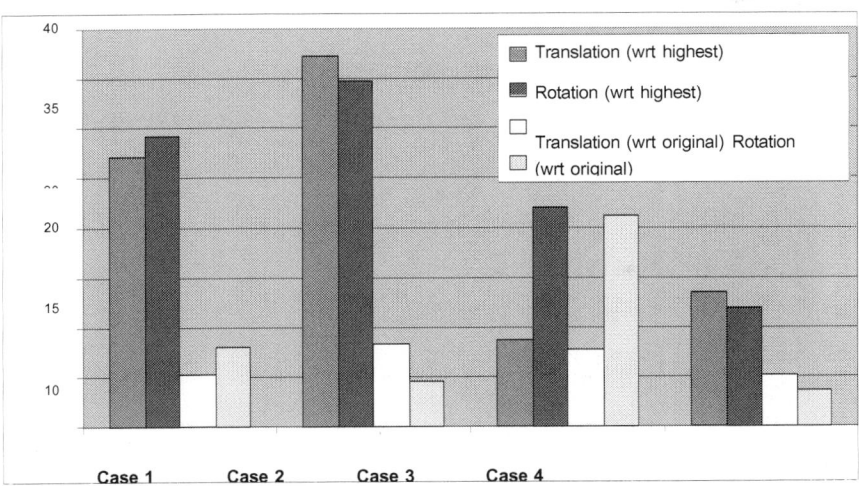

Figure. 11. Comparison of cycle time reduction percentage with respect to highest and original cycle time in four case studies

3. COMBINED DRIVE-TRAIN AND ROBOT PLACEMENT OPTIMIZATION

3.1 Research background

Offline programming of industrial robots and simulation-based robotic work cell design have become an increasing important approach for the robotic cell designers. However, current robot programming systems do not usually provide functionality for finding the optimum task placement within the workspace of a robot manipulator (or relative placement of working stations and robots in a robotic cell). This poses two principal challenges: 1) Develop methodology and algorithms for formulating and solving this type of problems as optimization problems and 2) Implement such methodology and algorithms in available engineering tools for robotic cell design engineers.

In the past years, much research has been devoted to the methodology and algorithm development for solving optimization problem of designing robotic work cells. In Section 2, a robust and sophisticated approach for optimal task placement problem has been proposed, developed, and implemented in one of the well-known robot offline programming tool RobotStudio from ABB. In this approach, the cycle time is used as the objective function and the goal of the task placement optimization is to place a pre-defined task defined in a robot motion path in the workspace of the robot to ensure minimum cycle time.

In this section, firstly, the task placement optimization problem discussed in Section 2 will be extended to a multi-objective optimization problem formulation. Design space for exploring the trade-offs between cycle time performance and lifetime of some critical drive- train component as well as between cycle time performance and total motor power consumption are presented explicitly using multi-objective optimization. Secondly, a combined task placement and drive-train optimization (combined optimization will be termed in following texts throughout this chapter) will be proposed using the same multi- objective optimization problem formulation. To authors' best knowledge, very few literature has disclosed any previous research efforts in these two types of problems mentioned above.

3.2 Problem statement

Performance of a robot may be modified by re-setting robot drive-train configuration parameters without any need of modification of hardware of the robot. Performance of a robot depends on positioning of a task that the robot performs in the workspace of the robot. Performance of a robot may therefore be optimized by either optimizing drive-train of the robot (Pettersson, 2008; Pettersson & Ölvander, 2009; Feng et al., 2007) or by optimizing positioning of a task to be performed by the robot (Kamrani et al., 2009).

Two problems will be investigated: 1) Can the task placement optimization problem described in Section 2 be extended to a multi-objective optimization problem by including both cycle time performance and lifetime of some critical drive-train component in the objective function and 2) What significance can be expected if a combined optimization of a robot drive-train and robot task positioning (**simultaneously optimize a robot drive-train and task positioning**) is conducted by using the same multi-objective optimization problem formulation.

In the first problem, additional aspects should be investigated and quantified. These aspects include 1) How to formulate multi-objective function including cycle time performance and lifetime of critical drive-train component; 2) How to present trade-off between the conflicting objectives; 3) Is it feasible and how efficient the optimization problem may be solved; and 4) How the solution space would look like for the cycle time performance vs. total motor power consumption.

In the second problem investigation, in addition to those listed in the problem formulation for the first type of problem discussed above, following aspects should be investigated and quantified: 1) Is it meaningful to conduct the combined optimization? A careful benchmark work is requested; 2) How efficient the optimization problem may be solved when additional drive-train design parameters are included in the optimization problem? Will it be applicable in engineering practice?

It should be noted that, focus of this work presented in Section 3 is on methodology development and validation. Therefore implementation of the developed methodology is not included and discussed. However, the problem and challenge for future implementation of the developed methodology for the combined optimization will be clarified.

3.3 Methodology

3.3.1 Robot performance simulation

A special version of the ABB virtual controller is employed in this work. It allows access to all necessary information, such as motor and gear torque, motor and gear speed, for design use. Based on the information, total motor power consumption and lifetime of gearboxes may be calculated for used robot motion cycle. The total motor power is calculated by summation of power of all motors present in an industrial robot. The individual motor power consumption is calculated by sum of multiplication of motor torque and speed at each simulation time step. The lifetime of gearbox is calculated based on analytical formula normally provided by gearbox suppliers.

3.3.2 Objective function formulation

The task placement optimization has been formulated as a multi-objective design optimization problem. The problem is expressed by

$$\min F(\boldsymbol{DV}) = w_1 \times CT_{norm}(\boldsymbol{DV}) + w_2 \times 1/LT_{norm}(\boldsymbol{DV})$$

(5)

where CT_{norm} is a normalzied cycle time, calculated by

$$CT_{norm} = CT/CT_{original}$$

(6)

CT is the cycle time at each function evaluation in the optimization loop. $CT_{original}$ is the cycle time of the robot motion cycle with original task placement and original drive-train parameter setup for combined optimization. LT_{norm} is a normalized lifetime of gearbox of some selected critical axis. It is calculated by

$$LT_{norm} = LT/LT_{original} \tag{7}$$

LT is the lifetime of some critical gearbox selected based on the actual usage of the robot at each function evaluation in the optimization loop. $LT_{original}$ is the lifetime of the selected gearbox of the robot motion cycle with original task placement and original drive-train parameter setup for combined optimization. W_1 and W_2 are two weighting factors employed in the weighted-sum approach for multi-objective optimization (Ölvander, 2001). DV is a design variable vector.

Two optimization case studies have been conducted. Robot task placement optimization with the design variable vector defined as

$$DV = [\Delta X, \Delta Y, \Delta Z]^T \tag{8}$$

and combined optimization with the design variable vector defined as

$$DV = [\Delta X, \Delta Y, \Delta Z, DV_1, DV_2, ... DV_n]^T \tag{9}$$

where $DV_1, DV_2, DV_3, ... DV_n$ are the drive-train configuration parameters, while $\Delta X, \Delta Y, \Delta Z$ are the *change* in translational coordinates of all robot targets defining the position of a task.

3.3.3 Optimizer: ComplexRF

The optimization algorithm used in this work is the Complex method proposed by Box (Box, 1965). It is a non-gradient method specifically suitable for this type of simulation- based optimization. Figure 12 shows the principle of the algorithm for an optimization problem consisting of two design variables. The circles represent the contour of objective function values and the optimum is located in the center of the contour. The algorithm starts with randomly generating a set of design points (see the sub-figure titled "Start"). The number of the design points should be more than the number of design variables. The worst design point is replaced by a new and better design point by reflecting through the centroid of the remaining points in the complex (see the sub-figure titled "1. Step"). This procedure repeats until all design points in the complex have converged (see last two sub-figures from left). This method does not guarantee finding a global optimum. In this work, an improved version of the Complex, or normally referred to as ComplexRF, is used, in which a level of randomization and a forgetting factor are introduced for improvement of finding the global optimum (Krus et al., 1992; Ölvander, 2001).

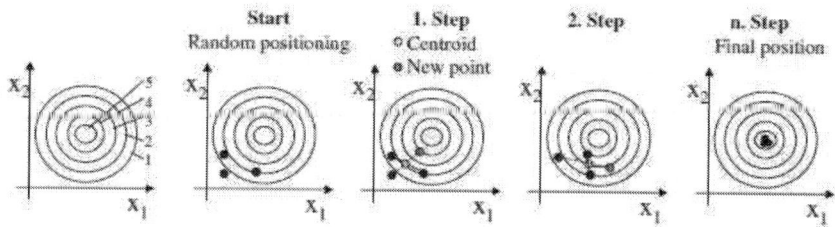

Figure. 12. The progress of the Complex method for a two dimensional example, with the optimum located in the center of the circles (Reprinted with permission from Dr. Johan Ölvander)

3.3.4 Workflow

The workflow of the proposed methodology starts with an optimizer generating a set of design variables. The variables defining robot task placement are used to manipulate the position of the robot task. The variables defining robot drive-train parameters are used to manipulate the drive-train parameters. The ABB robot motion simulation tool is run using the new task position and new drive-train setup parameters. Simulation results are used for computing objective function values. A convergence criterion is evaluated based on the objective function values. This optimization loop is terminated when either the optimization is converged or the limit for maximum number of function evaluations is reached. Otherwise, the optimizer analyzes the objective function values and proposes a new trial set of design variable values. The optimization loop continues until the convergence criterion is met.

3.4 Results on combined optimization

3.4.1 Case-I: Optimal robot usage for a spot welding application

In this case study, an ABB IRB6600-255-175 robot is used. The robot has a payload handling capacity of 175 kg and a reach of 2.55 m. A payload of 100 kg is defined in the robot motion cycle. The robot motion cycle used is a design cycle for spot welding application. The motion cycle consists of about 50 robot tool position targets. Maximum speed is programmed between any adjacent targets. A graphical illustration of the robot motion cycle is shown in Figure 5.

3.4.1.1 Task placement optimization

Only path translation is employed in the task placement optimization. Three design variables ΔX, ΔY, and ΔZ are used. They are added to all original robot targets so that the original placement of the robot task may be manipulated by ΔX in X coordinates, by ΔY in Y coordinates, and by ΔZ in Z corodinates. The limits for the path translation are

$$\Delta X \in (-0.1\ m, 0.1\ m)$$
$$\Delta Y \in (0\ m, 0.8\ m)$$
$$\Delta Z \in (-0.1\ m, 0.1\ m)$$

The weighting factors W_1 and W_2 in objective function (5) are set to $W_1 = 150$ and $W_1 = 100$ ⬛⬛⬛ in this task placement optimization.

The convergence curve of the task placement optimization is shown in Figure 13(a). The optimization is well converged after about 100 function evaluations. The total optimization time is about 15 min on a portable PC with Intel(R) Core(TM) 2 Duo CPU T9600 @ 2.8 GHz.

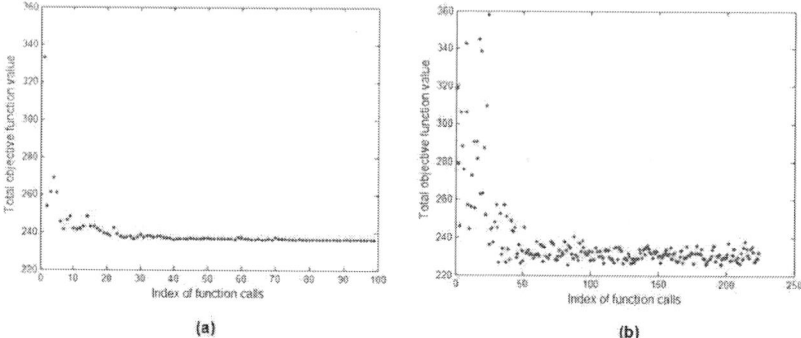

(a)   (b)

Figure. 13. Convergence curve. (a) for optimal task placement and (b) for combined optimization, (ABB IRB6600-255-175 robot)

Figure 14(a) shows the solution space of normalized lifetime of a critical gearbox as function of normalized cycle time. The cross symbol in blue color indicates the coordinate representing normalized lifetime and normalized cycle time obtained on the robot motion cycle programmed at original task placement. The results presented in the figure suggest one solution point with 8% reduction in cycle time (or improved cycle time performance) on the cost of about 50% reduction in the lifetime (point A1 in the figure 14(a)). Another interesting result disclosed in the figure is solution points in region A2, where about 20% increase in lifetime may be achieved with the same or rather similar cycle time performance. Figure 15(a) shows the solution space of normalized total motor power consumption as function of cycle time. The normalized total motor power consumption is obtained by actual total motor power consumption at each function evaluation in the optimization loop divided by the total motor power consumption obtained on the robot motion cycle programmed at original task placement. The cross symbol in blue color indicates the coordinate representing normalized total motor power consumption and cycle time obtained on the robot motion cycle programmed at original task placement. The results presented in the figure disclose that the ultimate performance improvement point suggested by point A1 in figure 14(a) results in an increase of about 20% in total motor power consumption (point B1 in the figure 15(a)). Another interesting result disclosed in the figure is solution points in region B2, where about 5% saving of total motor power consumption may be achieved for the solution points presented in region A2 in figure 14(a). In other words, the solution points in region A2 in figure 15(a) suggest not only increase in lifetime but also saving of total motor power consumption.

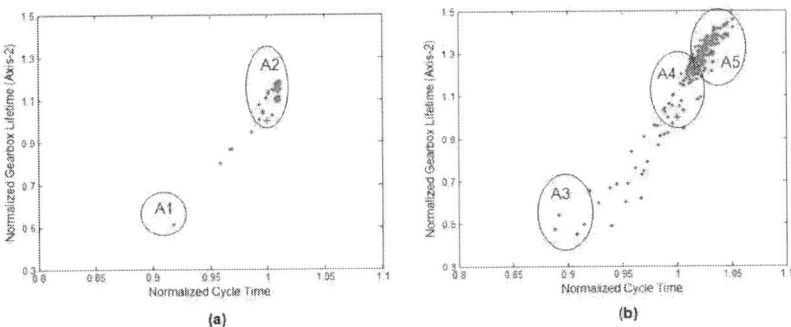

Figure. 14. Solution space of normalized lifetime of gearbox of axis-2 vs. normalized cycle time. (a) for optimal task placement and (b) for combined optimization, (ABB IRB6600-255-175 robot)

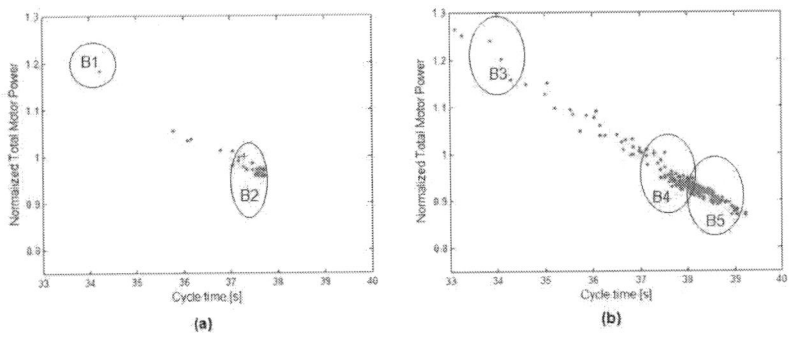

Figure. 15. Solution space of normalized total motor power vs. cycle time. (a) for optimal task placement and (b) for combined optimization, (ABB IRB6600-255-175 robot)

3.4.1.2 Combined task placement and drive-train optimization

The combined optimization involves both path translation for robot task placement and change of robot drive-train parameter setup. Two sets of design variables are used, the first set includes ΔX, ΔY, and ΔZ described in the task placement optimization; the second set includes nine design variables $DV_1, DV_2, ... DV_9$ which are scaling factors to be multiplied to the original drive-train parameters of the three main axes (axes 1-3). The limits for the path translation are the same as those used in the task placement optimization, i.e., the same as in (10).

The limits for the $DV_1, DV_2, ... DV_9$ are

$$DV_i \in (0.9, 1.2), where \; i = 1, 2 ..., 9$$

(11)

The weighting factors W_1 and W_2 are also set to $w_1 = 150$ and $w_2 = 100$ in this combined optimization.

To ease the benchmark work of task placement optimization and the combined optimization, the results of the combined optimization are presented in the same figures as those of task placement optimization. In addition, the figures are carefully prepared at the same scale.

Figure 13(b) shows the convergence curve of the combined optimization. The maximum limit of function evaluations for the optimizer is set to be 225. Optimization is interrupted after the maximum number of function evaluation limit is reached. The total optimization time is about 45 min on the same portable PC used in this work.

Figure 14(b) shows the solution space of normalized lifetime of the same critical gearbox as function of normalized cycle time. The cross symbol in blue color indicates the coordinate representing normalized lifetime and normalized cycle time obtained on the robot motion cycle programmed at original task placement and with original drive-train parameter setup values. The results presented in the figure suggest one solution point with more than 10% reduction in cycle time (or improved cycle time performance) on the cost of about 50% reduction in the lifetime (point A3 in the figure). Another result set disclosed in region A4 in the figure indicates up to 25% increase in lifetime that may be achieved with the same or rather similar cycle time performance. When a cycle time increase of up to 5% is allowed in practice, the lifetime of the critical gearbox may be increased by as much as close to 50% (region A5).

Figure 15(b) shows the solution space of normalized total motor power consumption as function of cycle time. The normalized total motor power consumption is obtained by actual total motor power consumption at each function evaluation in the optimization loop divided by the total motor power consumption obtained on the robot motion cycle programmed at original task placement and with original drive-train parameter setup values. The cross symbol in blue color indicates the coordinate representing normalized total motor power consumption and cycle time obtained on the robot motion cycle programmed at original task placement and with original drive-train parameter setup values. The results presented in the figure disclose that the ultimate performance improvement point suggested by point A3 in figure 14(b) results in an increase of about 20% in total motor power consumption (point B3 in the figure 15(b)). Another interesting result set disclosed in the figure is solution points in region B4, where about 5% saving of total motor power consumption may be achieved for the solution points presented in region A4 in figure 14(b). In other words, the solution points in region A4 in figure 14(b) suggest not only increase in lifetime but also saving of total motor power consumption. When a cycle time increase of up to 5% is allowed, not only the lifetime of the critical gearbox may be increased by as much as close to 50% (region A5) but also the total motor power consumption may be reduced by more than 10%.

3.4.1.3 Comparison between task placement optimization and combined optimization

When comparing the task placement optimization with combined optimization, it is evident that the combined optimization results in much large solution space. This implies in practice that robot cell design engineers would have more flexibility to place the task and setup drive-train parameters in more optimal way. However, the convergence time is also longer, due to the increase in number of design variables introduced in the combined optimization. In addition, changing drive-train parameters in robot cell optimization may pose additional consideration in robot design, so that the adaptation of drive-train in cell optimization would not result in unexpected consequence for a robot manipulator.

3.4.2 Case-II: Optimal robot usage for a typical material handling application

In this case study, an ABB IRB6640-255-180 robot is used. The robot has a payload handling capacity of 180 kg and a reach of 2.55 m. The payload used in the study is 80 kg. The robot motion cycle used is a typical pick-and-place cycle with 400 mm vertical upwards - 2000mm horizontal - 400mm vertical downwards movements – then reverse trajectory to return to the original position. Maximum speed is programmed between any adjacent targets.

3.4.2.1 Task placement optimization

Only path translation is employed in the task placement optimization. Three design variables, ΔX, ΔY, and ΔZ are used to manipulate the task position in the same manner as discussed in the Case-I. The limits for the path translation are

$$\Delta X \in (-0.1\ m, 0.1\ m)$$
$$\Delta Y \in (-0.1\ m, 0.1\ m)$$
$$\Delta Z \in (-0.2\ m, 0.5\ m)$$

(12)

The weighting factors w_1 and W_2 are set to $w_1 = 100$ and $w_2 = 100$ in this task placement optimization.

The convergence curve of the task placement optimization is shown in Figure 16(a). The optimization is converged after 290 function evaluations. The total optimization time is about 40 min on the same portable PC used in this work.

Figure 17(a) shows the solution space of normalized lifetime of a critical gearbox as function of normalized cycle time. The cross symbol in blue color indicates the coordinate representing normalized lifetime and normalized cycle time obtained on the robot motion cycle programmed at original task placement. The results presented in the figure suggest one set of solution points with close to 6% reduction in cycle time (or improved cycle time performance) with somehow improved lifetime of the critical axis under study (region A6 in the figure). Another interesting result set disclosed in the figure is solution points in region A7, where about 20% increase in lifetime may be achieved with 3-4% improvement of cycle time performance. In engineering practice, 3-4% cycle time improvement can imply rather drastic economic impacts.

Figure 18(a) shows the solution space of normalized total motor power consumption as function of cycle time. The cross symbol in blue color indicates the coordinate representing normalized total motor power consumption and cycle time obtained on the robot motion cycle programmed at original task placement. The results presented in the figure disclose.

that the solution points with more than 4% cycle time performance improvement (region B6) result in at least 20% increase in total motor power consumption.

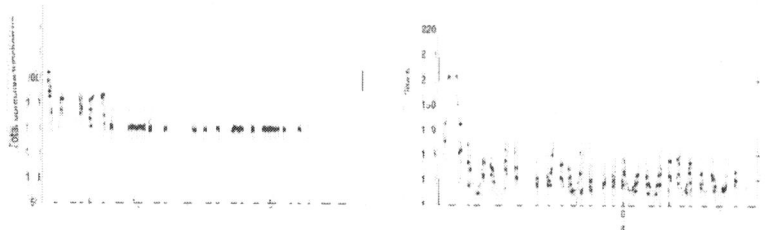

Figure. 16. Convergence curve. (a) for optimal task placement and (b) for combined optimization, (ABB IRB6640-255-180 robot)

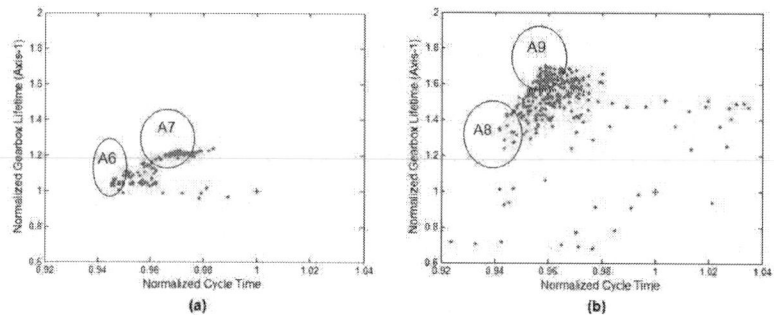

Figure. 17. Solution space of normalized lifetime of gearbox of axis-2 vs. normalized cycle time. (a) for optimal task placement and (b) for combined optimization, (ABB IRB6640-255-180 robot)

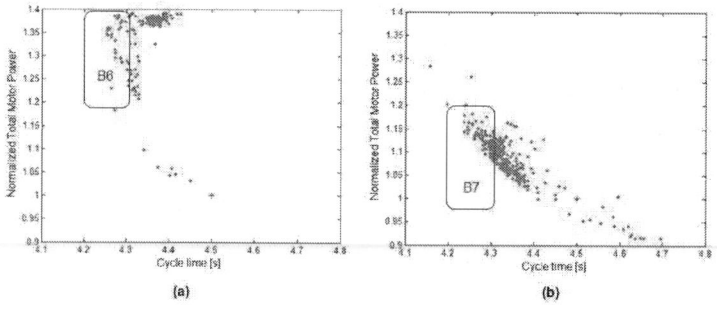

Figure. 18. Solution space of normalized total motor power vs. cycle time. (a) for optimal task placement and (b) for combined optimization, (ABB IRB6640-255-180 robot)

3.4.2.2 Combined task placement and drive-train optimization

As discussed in Case-I, the combined optimization involves both path translation for robot task placement and change of robot drive-train parameter setup. The same two sets of design variables are used. The limits for the path translation are the same as those used in the task placement optimization which are defined by (13).

The limits for the $DV_1, DV_2, ... DV_9$ are

$$DV_i \in (0.9, 1.2), where\ i = 1, 2 \dots, 9 \tag{13}$$

The weighting factors W_1 and W_2 are also set to $W_1 = 100$ and $W_2 = 100$ in this combined optimization.

For the same reason, the results of the combined optimization are presented in the same figures as those of task placement optimization. In addition, the figures are carefully prepared at the same scale.

Figure 16(b) shows the convergence curve of the combined optimization. The maximum limit of function evaluations for the optimizer is set to be 325. Optimization is interrupted after the maximum number of function evaluation limit is reached. The total optimization time is about 65 min on the same portable PC used in this work.

Figure 17(b) shows the solution space of normalized lifetime of the same critical gearbox as function of normalized cycle time. The cross symbol in blue color indicates the coordinate representing normalized lifetime and normalized cycle time obtained on the robot motion cycle programmed at original task placement and with original drive-train parameter setup values. The results presented in region A8 in the figure suggest a set of solution points with close to 6% reduction in cycle time but with clearly more than 20% increase in the lifetime. Another result set disclosed in region A9 in the figure indicates more than 60% increase in lifetime and with 3-4% improved cycle time performance!

Figure 18(b) shows the solution space of normalized total motor power consumption as function of cycle time. The cross symbol in blue color indicates the coordinate representing normalized total motor power consumption and cycle time obtained on the robot motion cycle programmed at original task placement and with original drive-train parameter setup values. The solution points disclosed in region B9 indicate that the solution points with more than 4% cycle time performance improvement result in maximum 20% increase in total motor power consumption.

3.4.2.3 Comparison between task placement optimization and combined optimization

Compared to task placement optimization, it is evident that the combined optimization results in much large solution space. This implies in practice that robot cell design engineers would have more flexibility to place the task and setup drive-train parameters in more optimal way. Even more significantly, the optimization results obtained on this typical pick- and-place cycle reveals more interesting observations. When the same cycle time improvement may be achieved, much more significant lifetime improvement may be achieved by combined optimization and the same is true for the total motor power consumption.

However, the convergence time is longer and optimization has to be interrupted using pre- defined maximum number of function evaluations, due to the increase in number of design variables introduced in the combined optimization. In addition, the same consequence is evident: changing drive-train parameters in robot cell optimization may pose additional consideration in robot design, so that the adaptation of drive-train in cell optimization would not result in unexpected consequence for a robot manipulator.

3.5 Summary of the Results of Section 3

Multi-objective robot task placement optimization shows obvious advantage to understand the trade-off between cycle time performance and lifetime of critical drive-train component. Sometimes, it may be observed that the cycle time performance and lifetime can be simultaneously improved. When task placement optimization involving only path translation is conducted, reasonable optimization time can be achieved.

The combined optimization of a robot drive-train and robot task placement, in comparison with task placement optimization, has disclosed even more advantages in achieving 1) wider solution space and 2) even more simultaneously improved cycle time performance and lifetime. Benefit of the combined optimization has been evident. Even though the optimization time can be nearly 2-3 times longer than task placement optimization, it can still be justified to be used in engineering practice; namely, earning from longer lifetime of a robot installation is greater than the calculation costs.Furthermore, this suggests that more efforts should be devoted in the future to; 1) better understanding of the multi-objective combined optimization problem and its impact on simulation-based robot cell design optimization; 2) improving efficiency of the optimization algorithms; 3) including collision-free task placement; and finally 4) sophisticated software implementation for engineering usage.

The plots of lifetime of critical component as function of cycle time performance and that of total motor power consumption as function of cycle time performance are also suggested in this work. This graphical representation of the solution space can further ease robot cell design engineers to better understand the trade-off between lifetime of critical drive-train component or total motor power consumption to cycle time performance and therefore choose better design solution that meets their goal.

4. CONCLUSIONS AND OUTLOOK

4.1 Single Objective Optimization

The results confirm that the problem of path placement in a robot work cell is an important issue in terms of manipulator cycle time. Cycle time greatly depends on the path position relative to the robot manipulator. Up to the 37.2% variation of cycle time has been observed which is remarkably high. In other words, the cycle time is very sensitive to the path placement. Algorithm and tool were developed to determine the optimal robot position by path translation and path rotation approaches. Several case studies were considered to evaluate and verify the developed tool for optimizing the robot position in a robotic work cell. Results disclose that an increase in productivity up to 37.2% can be achieved which is profoundly valuable in industrial robot application. Therefore, using this tool can significantly benefit the companies which have similar manipulators in use.

It is certain that employing this methodology has many important advantages. First, the cycle time reduces significantly and, therefore, the productivity increases. The method is easy to implement and the expense is only simulation cost, i.e., not any extra equipment is needed to be designed or purchased. The solution coverage is considerably broad, meaning that any type of robots and paths can be optimized with the proposed methodology. Another merit of the algorithm is that convergence is not an issue, i.e., reducing the cycle time can be assured. However, a disadvantage is that a global optimum cannot be guaranteed. The importance of the developed methodology is not confined only to the robot end-user application. Robot designers can also take advantage of the proposed methodology by optimizing the robot parameters such as robot structure and drive-train parameters to improve robot performance. As a design application example, the idea of optimum relative position of robot and path can be applied to the design of a tool such as welding device or glue gun which is erected on the mounting flange of the robot. The geometry of the tool can be optimized by studying design parameters to achieve shorter cycle time. Another possibility can be to use the developed methodology for optimal robot placement to realize other optimization objective in robots such as minimizing the torque, energy consumption, and component wear.

One interesting issue that can be investigated is to consider the general problem of finding the optimum by translation and rotation of the path simultaneously. What has been demonstrated in section 2 of the current chapter is to find the optimum path location by either translation or rotation of the path. Obviously, it is also possible to apply both these approaches at the same time. This would probably further shorten the cycle time in comparison to the case when only one approach is used. However, developing an optimal strategy for concurrently applying both approaches is an interesting challenge for future research.

Another important subject to be investigated is to take into account constraints for avoiding collisions. In a real application, a robot is not alone in the work cell as other cell equipments can exist in the workspace of the robot. Hence, in real robot application it is important to avoid collision.

4.2 Multi-Objective Optimization

It is noteworthy that although the methodology is implemented in RobotStudio, the algorithm is general and not dependent on RobotStudio. Therefore, the same methodology and algorithm can be implemented in any other robotic simulation software for achieving time optimality.

Multi-objective robot task placement optimization shows obvious advantage to understand the trade-off between cycle time performance and lifetime of critical drive-train components. The combined optimization of a robot drive-train and robot task placement, in comparison with task placement optimization, discloses even more advantages in achieving wider solution space and even more simultaneously improved cycle time performance and lifetime.

However, weighted-sum approach for formulating the multi-objective function has experienced difficulties in this work, since the weighting factors have been observed to significantly affect the final solution. Hence, an advanced formulation of multi-objective function and algorithms for multi-objective optimization need to be investigThe relativeated.

In combined optimization, the reachability is presumed to be satisfied as the purpose of this work is to rather explore the effect and feasibility of the method. Nevertheless, advanced and practical solutions exist for reachability checking that need to be implemented in the future work. In this study, while the task placement defined in a robot program is manipulated, the relative placements among sub-tasks (representing in practice the relative placements among different robotic stations in a robot cell) are kept unchanged. In the future work, relative placements of sub-tasks in a robot cell can also be optimized using the proposed methodologies.

REFERENCES

1. Barral, D. & Perrin. J-P. & Dombre, E. & Lie'geois, A. (1999). Development of optimization tools in the context of an industrial robotic CAD software product, *International Journal of Advvanced Manufacturing Technology,* Vol. 15(11), pp. 822–831,

2. Box, G.E.P. & Hunter, W.G. & Hunter, J.S. (1978). Statistics for experimenters: an introduction to design, data analysis and model building, Wiley, New York

3. Box, M. J., (1965). A New Method of Constrained Optimization and a Comparison with Other Methods, *Computer Journal,* Vol 8, pp. 42-52

4. Fardanesh, B. & Rastegar, J. (1988). Minimum cycle time location of a task in the workspace of a robot arm, *Proceeding of the IEEE 23rd Conference on Decision and Control,* pp. 2280–2283

5. Feng, X. & Sander, S.T. & Ölvander, J. (2007). Cycle-based Robot Drive Train Optimization Utilizing SVD Analysis, *Proceedings of the ASME Design Automation Conference,* Las Vegas, September 4-7, 2007

6. Haug, E.J. (1992). Intermediate dynamics, Prentice-Hall, Englewood Cliffs, NJ

7. Kamrani, B. & Berbyuk, V. & Wäppling, D. & Stickelmann, U. & Feng, X. (2009). Optimal Robot Placement Using Response Surface Method, *International Journal of Advvanced Manufacturing Technology*, Vol. 44, pp. 201-210

8 . Khuri, A.I. & Cornell, J.A. (1987). Response surfaces design and analyses, Dekker, New York

9. Krus, P. & Jansson, A. & Palmberg, J-O. (1992). Optimization Based on Simulation for Design of Fluid Power Systems, *Proceedings of ASME Winter Annual Meeting*, Anaheim, USA

10. Luenberger, D.G. (1969). Optimization by vector space methods, Wiley, New York

11. Myers, R.H. & Montgomery, D. (1995). Response surface methodology: process and product optimization using designed experiments, Wiley, New York

12. Nelson, B. & Donath, M. (1990). Optimizing the location of assembly tasks in a manipulator's workspace, Journal of Robotic Systems, Vol 7(6), pp. 791–811,

13. Pettersson, M. & Ölvander, J. (2009). Drive Train Optimization for Industrial Robots, *IEEE Transactions on Robotics*, to be published

14. Pettersson, M. (2008). A PhD Dissertation, Linköping University, Linköping, Sweden Tsai, L.W. (1999). Robot analysis, Wiley, New York

15. Tsai, M.J. (1986). Workspace geometric characterization and manipulability of industrial robot. Ph.D. Thesis, Department of Mechanical Engineering, Ohio State University

16. Vukobratovic, M. (2002). Beginning of robotics as a separate discipline of technical sciences and some fundamental results—a personal view, *Robotica*, Vol. 20(2), pp. 223–235

1 7 .   Yoshikawa, T. (1985). Manipulability and redundancy control of robotic mechanisms, *Proceeding of the IEEE Conference on Robotics and Automation*, pp 1004–1009, St. Louis

18. Ölvander J. (2001). Multiobjective Optimization in Engineering Design – Applications to Fluid Power Systems, A PhD Dissertation, No. 675 at Linköping University

CHAPTER 4

Joint Torque Reduction of a Three Dimensional Redundant Planar Manipulator

Samer Yahya [1,*], Mahmoud Moghavvemi [1,2]
and Haider Abbas F. Almurib [3]

[1] Center of Research in Applied Electronics (CRAE), University of Malaya, Kuala Lumpur 50603, Malaysia

[2] Faculty of Electrical and Computer Engineering, University of Tehran, P.O. Box 14399-57131, Tehran, Iran

[3] Department of Electrical & Electronic Engineering, University of Nottingham Malaysia, Jalan Broga, Semenyih 43500, Malaysia

ABSTRACT

Research on joint torque reduction in robot manipulators has received considerable attention in recent years. Minimizing the computational complexity of torque optimization and the ability to calculate the magnitude of the joint torque accurately will result in a safe operation without overloading the joint actuators. This paper presents a mechanical design for a three dimensional planar redundant manipulator with the advantage of the reduction in the number of motors needed to control the joint angle, leading to a decrease in the weight of the manipulator. Many efforts have been focused on decreasing the weight of manipulators, such as using lightweight joints design or setting the actuators at the base of the manipulator and using tendons for the transmission of power to these joints. By using the design of this paper, only three motors are needed to control any n degrees of freedom in a three dimensional planar redundant manipulator instead of n motors. Therefore this design is very effective to decrease the weight of the manipulator as well as the number of motors needed to control the manipulator. In this paper, the torque of all the joints are calculated for the proposed manipulator (with three motors) and the conventional three dimensional planar manipulator (with one motor for each degree of freedom) to show the effectiveness of the proposed manipulator for decreasing the weight of the manipulator and minimizing driving joint torques.

KEYWORDS

redundant manipulator; dynamics; robot; rotary encoders; joint torques reduction

1. INTRODUCTION

Theoretically, for a structure of the robot manipulator one actuator can be mounted on each link to drive the next link via a speed reduction unit, but actuators and speed reducers installed on the distal end become the load for actuators installed on the proximal end of a manipulator, resulting in a bulky and heavy system [1]. To reduce the weight and the inertia of a robot manipulator, many mechanisms have been proposed so far to remove the weight restriction. Some reported by [2,3] include:

a.   Lightweight joint design based on a special rotary joint [4–6]

b.   Provision of a powerful slider at the base to bear as much required driving force as possible [7]

c.   The parallel mechanism is another method to reduce the mass and inertia of the manipulator [8]. A typical parallel manipulator consists of a moving platform that is connected with a fixed base by several limbs. Generally, the number of degrees of freedom of a parallel manipulator is equal to the number of its limbs. The actuators are usually mounted on or near the base, which contributes to reduce the inertia of manipulators, and

d. Concentration of the actuators at the base and transmission of the power to each joint through tendons or a special transmission mechanism [2,3,9]. This mechanism allows the actuators to be situated remotely on the manipulator base, allowing the manipulator to be made more lightweight and compact.

For a serial manipulator, direct kinematics are fairly straightforward, whereas inverse kinematics becomes very difficult. Reference [10] proposes a fused smart sensor network to estimate the forward kinematics of an industrial robot, while reference [11] measures the range data with respect to the robot base frame using the robot forward kinematics and the optical triangulation principle. The inverse kinematics problem is much more interesting and its solution is more useful, but one of the difficulties of inverse kinematics is that when a manipulator is redundant, it is anticipated that the inverse kinematics has an infinite number of solutions. This implies that, for a given location of the manipulator's end-effector, it is possible to induce a self-motion of the structure without changing the location of the end-effector. In this paper we depend on our prior works [12,13] which present a new method to solve the problem of multi-solutions of a three dimensinal planar redundant manipulator. Because this paper explains the dynamic of the manipulator and not its kinematics, the inverse kinematics methods will not be explained here. For more details about the inverse kinematics of redundant manipulators, our works [14–16] can be checked.

It is mentioned earlier that the proposed manipulator could be used to reduce the weight of the manipulator which yields to a decrease in the size (power) of the motors used to control the manipulator. To show the effectiveness of the proposed manipulator in reducing the torques of its motors the inverse dynamic of the manipulator has been calculated mathematically. The inverse dynamic model provides the joint torques in terms of the joint positions, velocities and accelerations. For robot design, the inverse dynamic model is used to compute the actuator torques, which are needed to achive a desired motion [17]. Several approaches have been proposed to model the dynamics of robots. The most frequently employed in robotics are the Lagrange formulation and the Newton-Euler formulation. Because the Lagrange formulation is conceptually simple and systematic [18], it has been used in this paper. The Lagrange formulation provides a description of the relationship between the joint actuator forces and the motion of the mechanism, and fundamentally operates on the kinetic and potential energy in the system [19].

The work presented in this paper is based on our previous work [14], which presents a mechanical design for a three dimensional planar redundant manipulator, which guarantees to decrease the weight of the manipulator by decreasing the number of motors needed to control it. Because the inverse kinematics model gives an infinite number of solutions for a redundant manipulator, consequently, secondary performance criteria can be optimized [17], such as avoiding singular configurations and minimizing driving joint torques. Reference [14] studied the kinematics of the manipulator of this paper and showed in details its ability to avoid singular configurations. A comparison of the manipulability index values and the manipulability ellipsoids for the manipulator is made with the manipulability index values and the manipulability ellipsoids of the PUMA arm to show the effectiveness of using the proposed manipulator to avoid singularity. In this paper, the dynamics of this manipulator are explained in detail. The contribution of this work is to explain the ability of this manipulator for joint torque minimization. The links and motors mass distribution is studied for both the proposed (with three motors) and conventional manipulators (six motors). The driving joint torques have been studied for the proposed manipulator for each joint and the results are compared with the results of the conventional manipulators to show the effectiveness of this manipulator for minimizing driving joint torques.

2. THE MECHANICAL DESIGN OF THE MANIPULATOR

To control the motion of the end-effector of the manipulator shown of Figure 1(a), all the motors of the manipulator should be controlled. For example, to control a five links planar redundant manipulator with the ability to rotate the entire manipulator around its vertical axis, the six motors (five motors for each joint angle and one motor to rotate the entire manipulator around its vertical axis) of the manipulator should be controlled. Using the method of our papers [12,13], the configuration of the manipulator will have three angles to be controlled instead of n angles. Figure 1(b) shows the configuration of the manipulator when there are just three angles that need to be controlled.

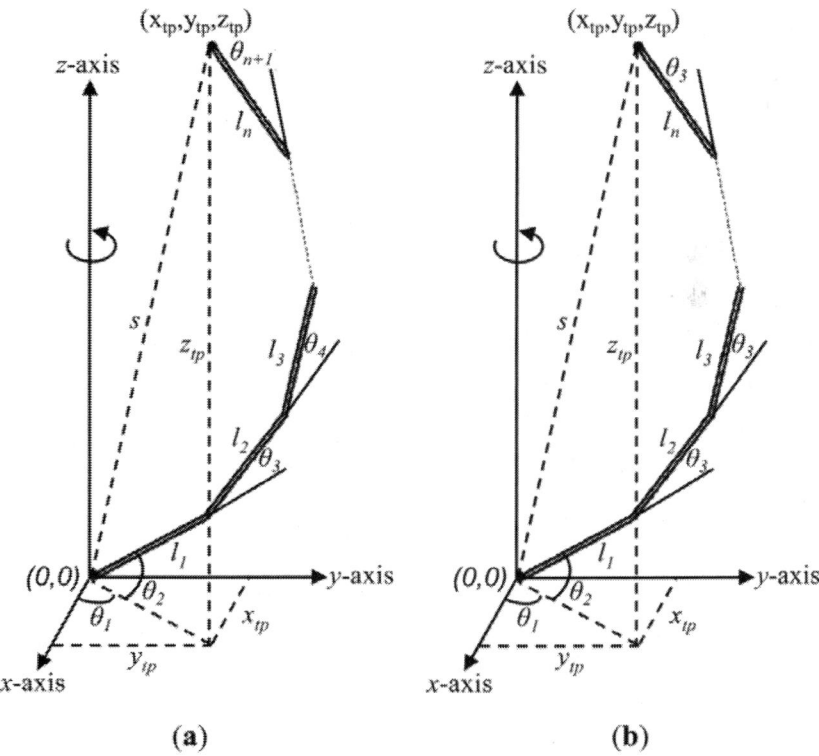

(a)   (b)

Figure 1. (a) A three dimensional planar redundant manipulator configuration; (b) A three dimensional planar redundant manipulator configuration using the method of [12,13].

Because the end-effector can follow any desired path by controlling three angles (θ_1, θ_2 and θ_3) only, therefore instead of using a motor for each joint angle, three motors can be used for controlling the manipulator. This means that for any number of degrees of freedom three dimensional planar redundant manipulators, the weight of the links will be significantly decreased using the proposed design. To make the manipulator capable of moving in a three dimensional work space, one motor will control the value of θ_1—this means controlling the rotation of the entire manipulator around the vertical axis. This motor is situated in such a way as to rotate the base of the manipulator around the z-axis. The second motor controls the value of θ_2, which means the rotation of the entire manipulator with its configuration. The motor is situated at the base. The third motor controls the value of θ_3 and this motor is situated on the first link. This motor will rotate the second link of manipulator about the second axis, and because all the next links should rotate about their axes by the same angle θ_3 therefore, there is no need to use motor for each joint angle, but the rotation of the second motor will be transferred to the next joints using gears boxes. Figure 2 shows the mechanism of the proposed manipulator.

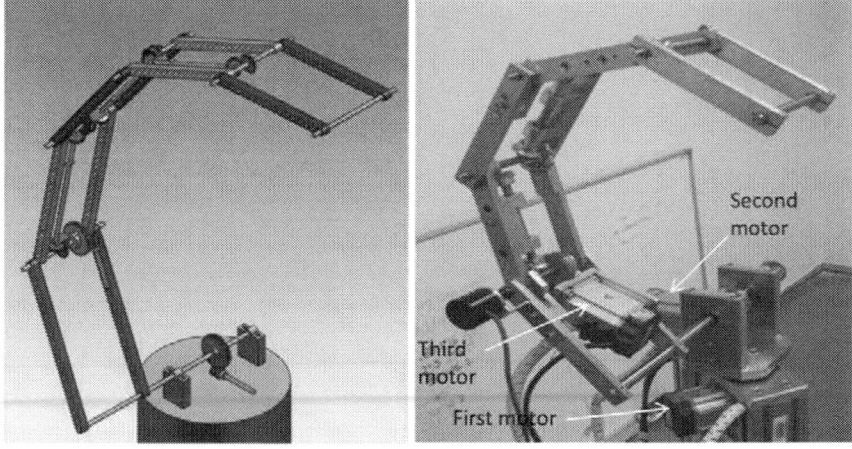

Figure 2. The manipulator used in experiments [14]. The draft of the manipulator using the SolidWorks software (**left**). The mechanical design of the manipulator (**right**).

Elaborating further, the second motor is connected to the first link using a worm gear to control the angle θ_2. Figure 3 shows the position of the second motor.

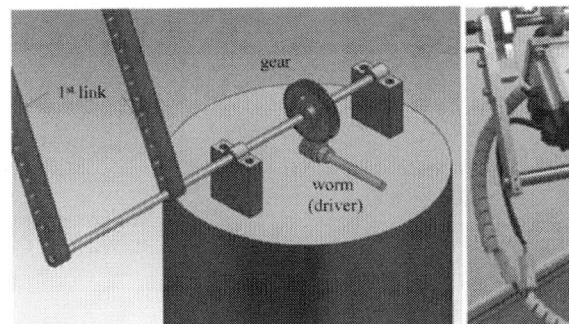

Figure 3. The design of the second joint angle (first link with second motor) of the manipulator [14]. The draft of the second joint angle using the SolidWorks software (**left**). The mechanical design of the second joint angle (**right**).

The third motor is connected to the second link using a worm gear for the same reasons it was used with the first link. Controlling the third motor means controlling the angle between the first link and the second link i.e., the angle θ_3. Figure 4 shows the position of the third motor.

The mechanism of the third link is shown in Figure 5. The same mechanism of the second link is used; the only one difference is that instead of using s worm as a driver and s wheel gear as a driven, two bevel gears are used. The same mechanism of the third link can be used with the next links. The last link has the mechanism shown in the Figure 6. For further details of the mechanical design of the manipulator, our reference [14] can be checked.

To ensure that all the links move at the same joint angle, the ratio between the bevel gears of each planetary gear should be equal to one. This means the bevel gears of each planetary gear should have the same diameter and number of teeth. If this arm is fixed, we get:

$$\frac{w_1}{w_2} = -\frac{N_2}{N_1}$$

(1)

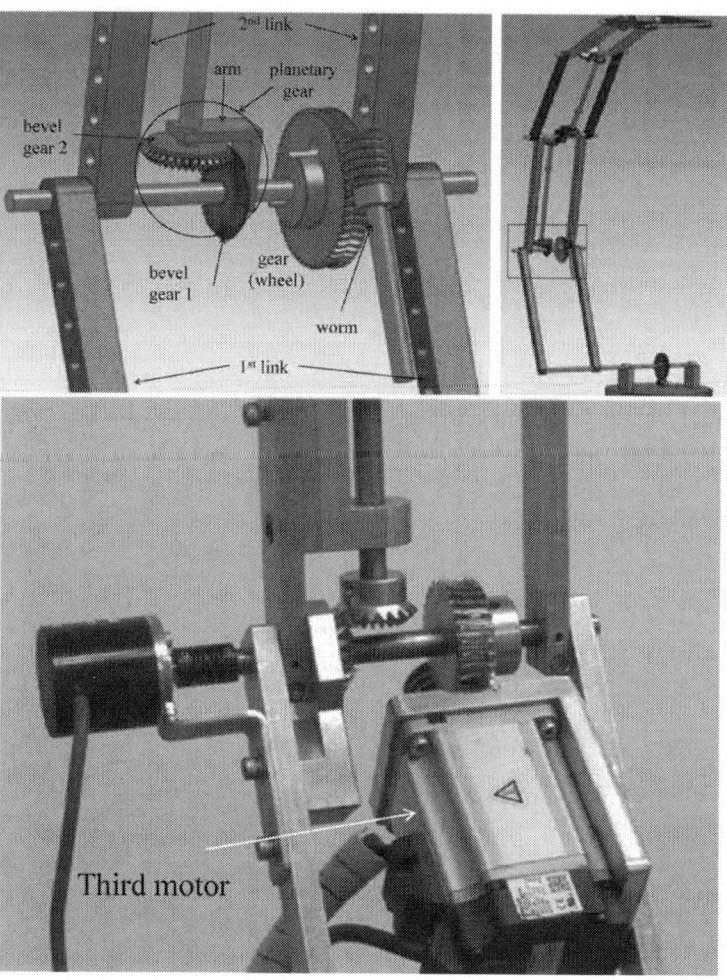

Figure 4. The design of the third joint angle (second link with third motor) of the manipulator [14]. The draft of the third joint angle using the SolidWorks software (**top left**). The draft of the whole manipulator using the SolidWorks software (**top right**). The mechanical design of the third joint angle (**bottom**).

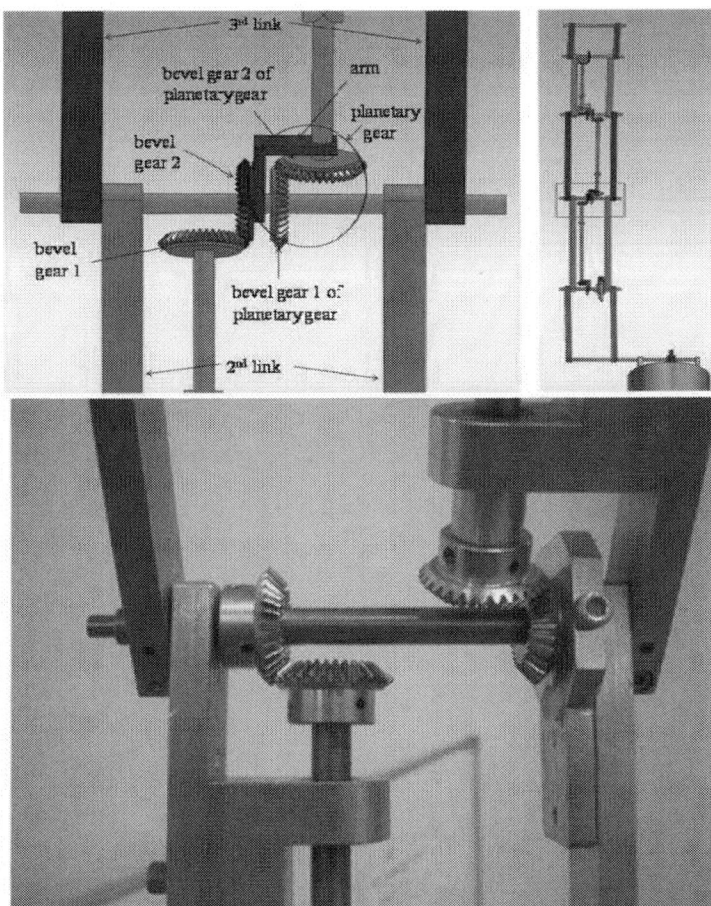

Figure 5. The design of the fourth joint angle (third link) of the manipulator [14]. The draft of the fourth joint angle using the SolidWorks software (**top left**). The draft of the whole manipulator using the SolidWorks software (**top right**). The mechanical design of the fourth joint angle (**bottom**).

where w is the angular velocity of gear and N is the number of teeth of gear. In our manipulator, it is noted that the first gear is fixed while the second gear and the arm are rotating. It is desired that both the arm and the second gear have the same angular velocity. Because the arm is not stationary, then we cannot use the previous equation. i.e., the mechanism is not an ordinary gear train but a planetary gear train. To convert this planetary gear train to an ordinary gear train, it is assumed that the arm is stationary while a first gear has an angular velocity and not fixed. This means that:

Figure 6. The last joint of the manipulator.

$$w_1' = w_1 - w_a \qquad (2)$$

$$w_a' = w_a - w_a = 0 \qquad (3)$$

And because the second gear will continue rotating with the same angular velocity, then:

$$w_2' = w_2 \qquad (4)$$

Now the Equation (1) can be rewritten as follows:

$$\frac{w_1'}{w_2'} = -\frac{N_2}{N_1} = \frac{w_1 - w_a}{w_2} \qquad (5)$$

For our manipulator it is desired to move both the arm and the second gear by the same angular velocity w which means:

$$\frac{(0)-w}{w} = -\frac{N_2}{N_1} \quad N_1 = N_2$$

(6)

To make the manipulator to have the ability to move in a three dimensional work space, a motor is added to the base of the manipulator to make the whole manipulator capable of rotating around the z-axis. This motor controls θ_1. Figure 7 shows the mechanism of the first motor.

Figure 7. The mechanism of the first motor.

To calculate the transformation matrix of the manipulator, the draft of the manipulator shown in Figure 8, is used. The corresponding link parameters of the manipulator are shown in Table 1. Where l_1, l_2, ..., l_5 are the length of the links, while d_1 is the offset between the origin and the end-effector.

From the links parameters shown in Table 1 and using Equation (7) which defines the transformation matrix T for the links [1], we compute the individual transformations for each link:

$$^{i-1}T_i = \begin{bmatrix} \cos\theta_i & -\sin\theta_i\cos\alpha_i & \sin\theta_i\sin\alpha_i & a_i\cos\theta_i \\ \sin\theta_i & \cos\theta_i\cos\alpha_i & -\cos\theta_i\sin\alpha_i & a_i\sin\theta_i \\ 0 & \sin\alpha_i & \cos\alpha_i & d_i \\ 0 & 0 & 0 & 1 \end{bmatrix}$$

(7)

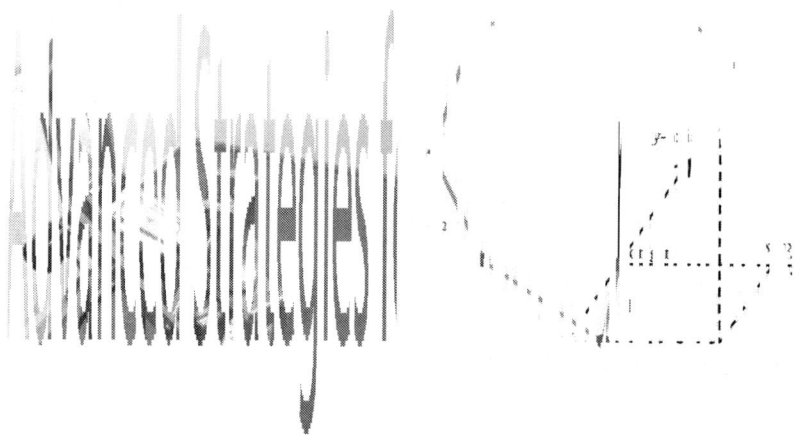

Figure 8. The manipulator used in experiments.

Table 1. Link parameters of the manipulator.

i	α	a	d	θ
1	90	0	0	θ_1
2	0	l_1	d_1	θ_2
3	0	l_2	0	θ_3
4	0	l_3	0	θ_4
5	0	l_4	0	θ_5
6	0	l_5	0	θ_6

where $c_i = cos(\theta_i)$ and $s_i = sin(\theta_i)$.

$$
{}^0T_1 = \begin{bmatrix} c_1 & 0 & s_1 & 0 \\ s_1 & 0 & -c_1 & 0 \\ 0 & 1 & 0 & 0 \\ 0 & 0 & 0 & 1 \end{bmatrix}, \ {}^1T_2 = \begin{bmatrix} c_2 & -s_2 & 0 & l_1 c_2 \\ s_2 & c_2 & 0 & l_1 s_2 \\ 0 & 0 & 1 & d_1 \\ 0 & 0 & 0 & 1 \end{bmatrix}, \ {}^2T_3 = \begin{bmatrix} c_3 & -s_3 & 0 & l_2 c_3 \\ s_3 & c_3 & 0 & l_2 s_3 \\ 0 & 0 & 1 & 0 \\ 0 & 0 & 0 & 1 \end{bmatrix}
$$

$$
{}^3T_4 = \begin{bmatrix} c_4 & -s_4 & 0 & l_3 c_4 \\ s_4 & c_4 & 0 & l_3 s_4 \\ 0 & 0 & 1 & 0 \\ 0 & 0 & 0 & 1 \end{bmatrix}, \ {}^4T_5 = \begin{bmatrix} c_5 & -s_5 & 0 & l_4 c_5 \\ s_5 & c_5 & 0 & l_4 s_5 \\ 0 & 0 & 1 & 0 \\ 0 & 0 & 0 & 1 \end{bmatrix}, \ {}^5T_6 = \begin{bmatrix} c_6 & -s_6 & 0 & l_5 c_6 \\ s_6 & c_6 & 0 & l_5 s_6 \\ 0 & 0 & 1 & 0 \\ 0 & 0 & 0 & 1 \end{bmatrix}
$$

$$(8)$$

Finally we obtain the product of all six link transforms:

$$^0T_6 = {}^0T_1 \, {}^1T_2 \, {}^2T_3 \, {}^3T_4 \, {}^4T_5 \, {}^5T_6$$

$$(9)$$

3. DYNAMICS OF THE MANIPULATOR

In this section, the torque of each joint is calculated. To show the effectiveness of the proposed manipulator, the joint torques are calculated using the proposed manipulator (using three motors only) and the conventional manipulators (a motor for each joint).

Let us assume for concreteness that the center of mass of each link is at its geometric center. For the manipulator used in our experiments, the mass of links without the motors are as follow: ml_1 = 760 gm, ml_2 = 720 gm, ml_3 = 680 gm, ml_4 = 640 gm, and finally ml_5 = 600 gm. These masses are calculated for the manipulator with l_1 = 19 cm, l_2 = 18 cm, l_3 = 17 cm, l_4 = 16 cm, l_5 = 15 cm and d_2 = 21 cm.

The mass of each motor is 1,500 gm; for the manipulator of the proposed design, the first motor and the second motor are located on the base and not on the links themselves. Therefore, for our manipulator, the mass of the first link will be equal to the mass of this link (760 gm) plus the mass of the motor (1,500 gm) that controls the next links. Because there are no more motors, the mass of the links will be: m_1 = 2,260 gm, m_2 = 720 gm, m_3 = 680 gm, m_4 = 640 gm, and m_5 = 600 gm. Figure 9(a) shows the mass of each link with its motor for the manipulator of the proposed design.

For the conventional three dimensional planar manipulator (one motor for each link), the mass of the first link will equal to the mass of link itself plus the mass of the motor which controls the second link position, i.e., 760 + 1,500 gm. The mass of the second link will equal to the mass of link itself plus the mass of the motor which controls the third link position, i.e., 720 + 1,500 gm. The mass of the third link will equal to the mass of third link plus the mass of the motor which controls the fourth link position, i.e., 680 +

1,500 gm. The mass of the fourth link will equal to the mass of fourth link plus the mass of the motor which controls the fifth link position, i.e., 640 + 1,500 gm, while the mass of the last link will equal to the mass of the link itself because there are no more motors, i.e., 600 gm. Figure 9(b) shows the mass of each link using the manipulator with five motors, while Table 2 shows the values of mass of the links using both the manipulator with two motors and the manipulator with five links.

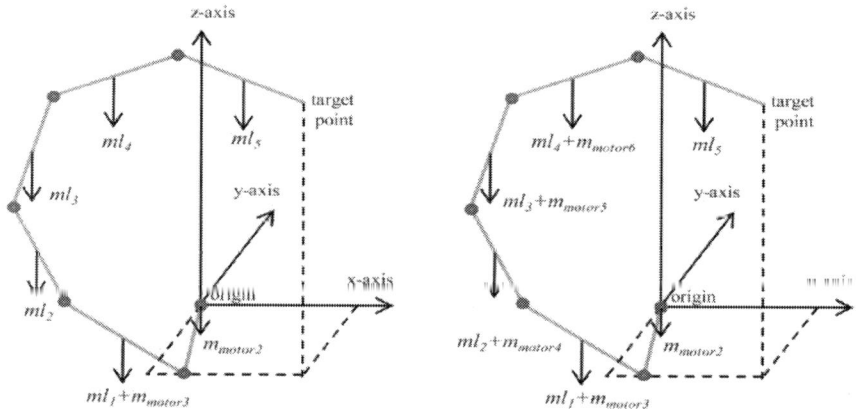

Figure 9. The position of mass for (**a**) the proposed manipulator; (**b**) the conventional manipulator.

Table 2. The mass of links for both the proposed and conventional manipulators.

m_n(gm)	Proposed manipulator	Conventional manipulator
m_1(gm)	1,500	1,500
m_2(gm)	2,260	2,260
m_3(gm)	720	2,220
m_4(gm)	680	2,180
m_5(gm)	640	2,140
m_6(gm)	600	600

It is clearly noted how the proposed method could be used to decrease the weight of manipulator. Decreasing the weight leads to a decrease of the torques of each link. The next section shows the results of the torques of each joint when the end-effector is following a desired path. Using the Lagrangian formulation, the dynamical equations of motion of the manipulator is:

$$\sum_{j=1}^{6} M_{ij}\ddot{q}_j + V_i + G_i = Q_i$$

$$(10)$$

for i = 1,2,....,6.

The first term in this equation is the inertia forces, the second term represents the Coriolis and centrifugal forces, and the third term gives the gravitational effects [1,20,21]. Dynamics equations of the manipulator are discussed in details in the Appendix.

As shown by dynamics equations, increasing the weight of motors will increase the torques needed to control the manipulator. In order to decrease the effect of the motors weight on the inertia of manipulators, parallel manipulators are used, as we mentioned earlier. For example in reference [22], the parallel manipulator is actuated by three servo-motors located at the base which contributes to reducing the inertia of manipulators. Reference [23] shows another way to decrease the effect of the motors weight on the inertia of manipulators. This reference shows a simple configuration design, which comprises of only three joints: two at the shoulder and one at the hand. In this design, the moment of inertia of the arm is constant and independent from the joint angles. In contrast for our manipulator, we see from Equations A21–A25 that the moment of inertia value is dependent on the joint angles.

4. SIMULATION RESULTS

This section shows the effectiveness of using the proposed manipulator to be used when it is desired to make the end-effector follow a desired path. This section has two examples. The first example calculates the torques using both manipulators (the proposed one and the conventional three dimensional planar manipulator) and shows how effective the proposed manipulator is in decreasing the torque of each joint required to move the manipulator. To verify the estimation results and compare between them and the results measured from the manipulator itself, the second example has been shown. This example shows the results if the torque using: (1) the conventional three dimensional planar manipulator with defined desired joint angles path, (2) the proposed manipulator with the defined desired joint angles path and finally (3) the proposed manipulator with the measured joint angles path when the joint angles follow the desired joint angles path.

Case One

Torque of each joint for both the manipulators is calculated to show the effectiveness of using the proposed manipulator to decrease the torque of each joint, Using the same manipulator with $1 - [19,18,17,16,15]^T$, and d_1 21 where all lengths are in cm, the joint angles path is defined as:

$$\theta_1(t) = -0.5\cos(4t)$$

(11)

$$\theta_2(t) = -\cos(2t) + 1$$

(12)

$$\theta_3(t) = -4\cos(t) + 3$$

(13)

It should be remembered that when using the proposed manipulator, θ_3, θ_4, θ_5, θ_6 are equal. To show the effectiveness of the proposed manipulator in decreasing the torque, Figure 10 shows the values of the torques of the first joint using both the manipulators, the proposed manipulator (with three motors) and the manipulator of six motors.

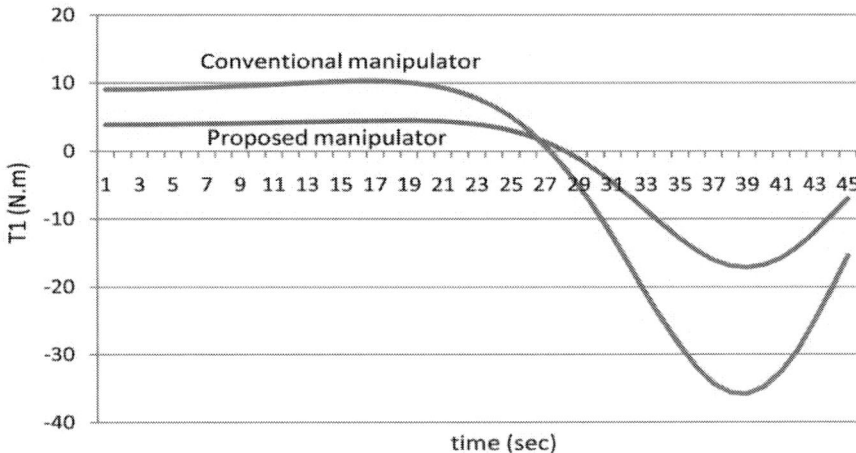

Figure 10. The values of the torques of the first joint using the both manipulators.

Figure 11 shows the values of the torques of the second joint using the manipulators.

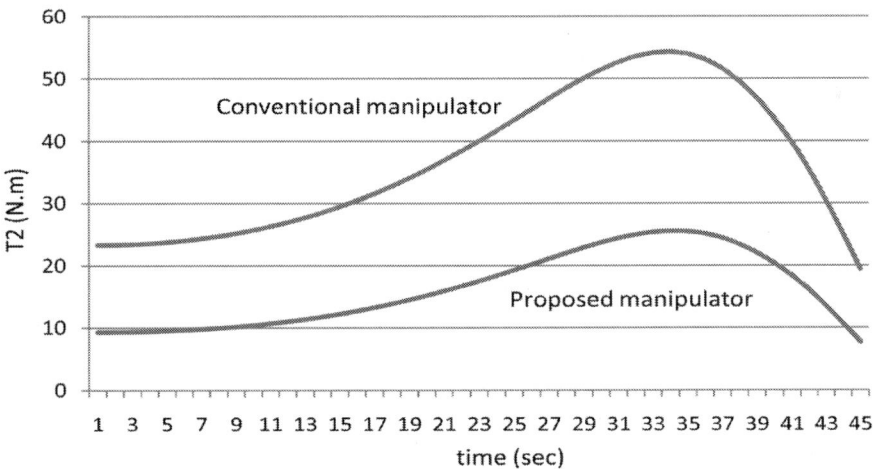

Figure 11. The values of the torques of the second joint using the both manipulators.

Figure 12 shows the absolute values of the torques of the third joint while Figure 13 shows the absolute values of the torques of the fourth joint using the both manipulators.

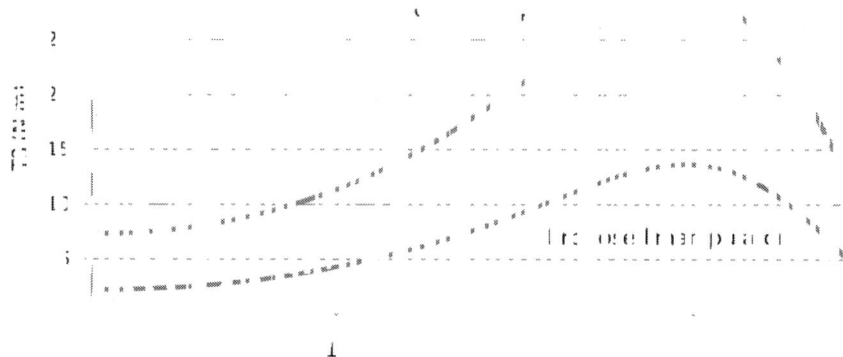

Figure 12. The values of the torques of the third joint using the both manipulators.

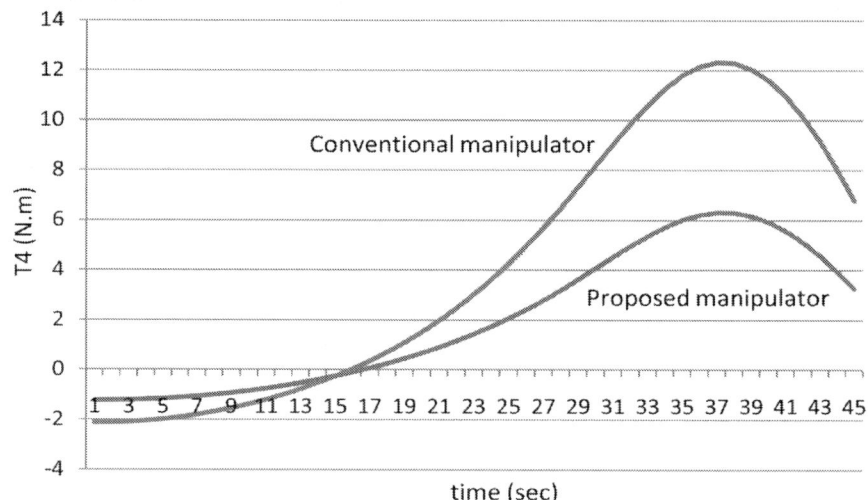

Figure 13. The values of the torques of the fourth joint using the both manipulators.

Figure 14 shows the torques of the fifth joint and finally Figure 15 shows the torques of the sixth joint angle using the both manipulators.

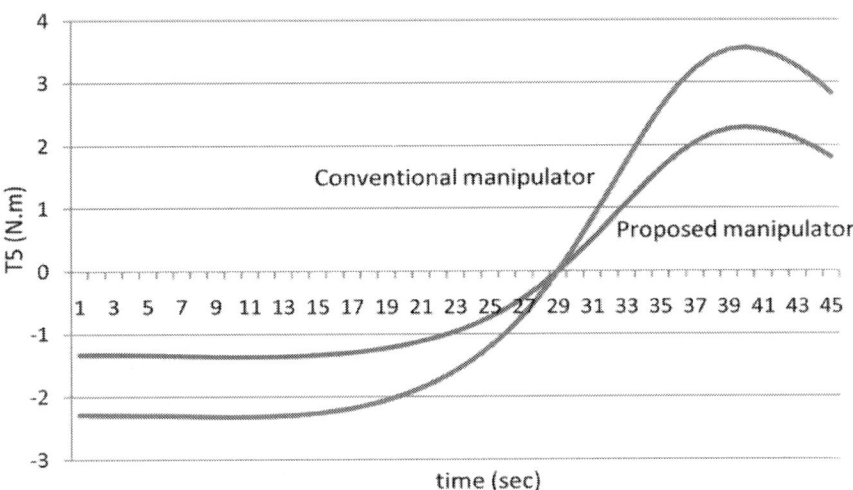

Figure 14. The values of the torques of the fifth joint using the both manipulators.

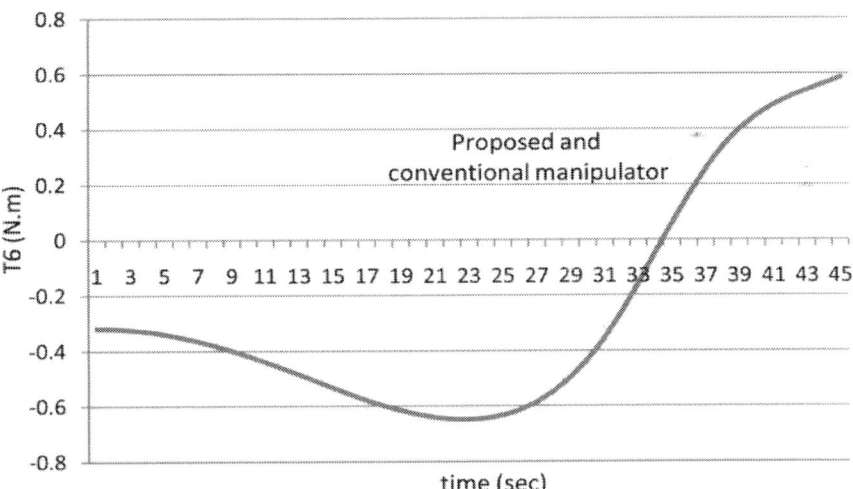

Figure 15. The values of the torques of the sixth joint using the both manipulators.

First of all, it is noted that the torque of the sixth joint has the same value using both the manipulators because the sixth link has the same mass for both the manipulators, in other words the mass of the sixth link is equal to the mass of the link itself only because it does not hold any motor.

Secondly, as mentioned earlier for the proposed manipulator, the third motor should balance the torque of all the third, fourth, fifth and the sixth joint. In other words, the torque of the third motor should equal to (T_3 + T_4 + T_5 + T_6) for the proposed manipulator. Figure 16 shows the power that the third motor should balance for both the manipulators. It is noted from this example that using the proposed manipulator not only decreases the number of motors used in the manipulator, but also decreases the torques of the motors used to control it.

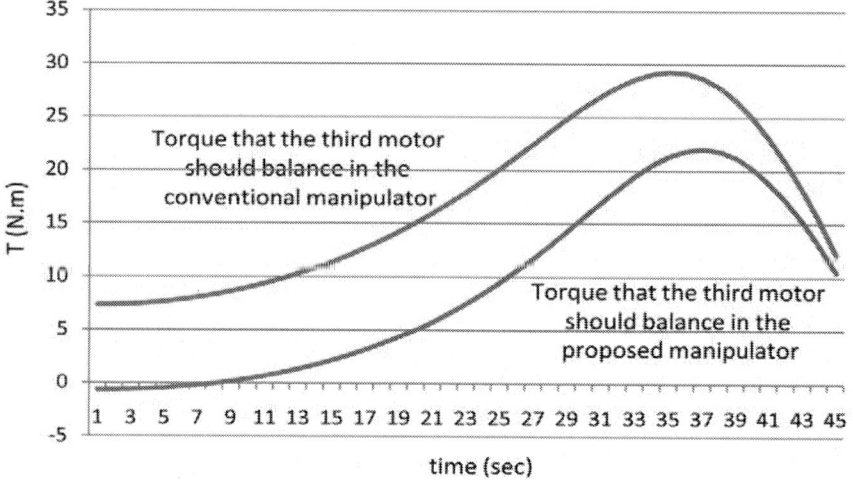

Figure 16. The values of the torques of the third motor using the both manipulators.

Case Two

The trajectory applied to robot in verification experiments in this case is:

$$\theta_1(t) = \frac{e^{0.7t}}{25}$$

(13)

$$\theta_2(t) = \frac{e^{0.5t}}{3}$$

(14)

$$\theta_3(t) = \frac{e^{0.9t}}{150}$$

(15)

Figure 17 shows the estimated (white) and measured (red) angle, angular velocity and angular acceleration of the first joint angle defined above. Figure 18 shows the estimated (white) and measured (red) angle, angular velocity and angular acceleration of the second joint angle.

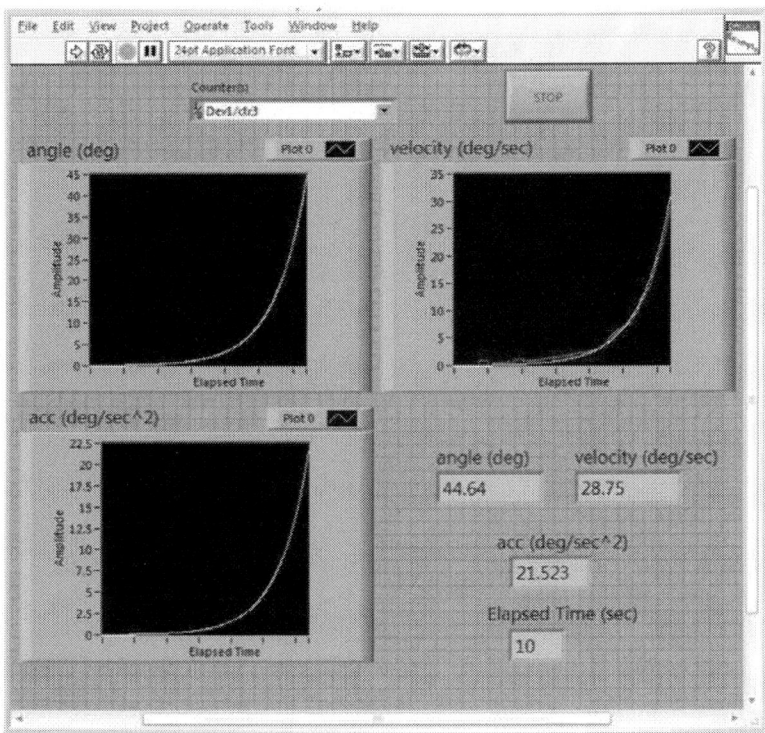

Figure 17. The values of the estimated and measured angular position, velocity and acceleration of the first joint angle (white: estimated, red: measured).

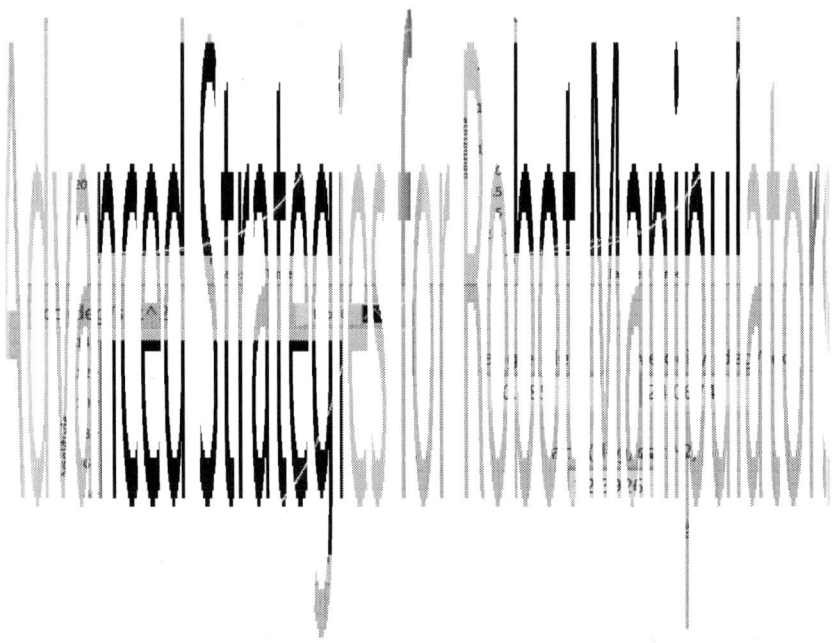

Figure 18. The values of the estimated and measured angular position, velocity and acceleration of the second joint angle (white: estimated, red: measured).

Figure 19 shows the estimated (white) and measured (red) angle, angular velocity and angular acceleration of the third joint angle of the manipulator. It should be remembered again that using the proposed manipulator, θ_3, θ_4, θ_5, θ_6 are equals.

Figures 20–25 show the comparison between the torque of each joint angle for: (1) the conventional three dimensional planar manipulator using the estimated joint angles path; (2) the proposed manipulator using the estimated joint angles path; and finally (3) the proposed manipulator using the measured angular position, velocity and acceleration of the manipulator joints.

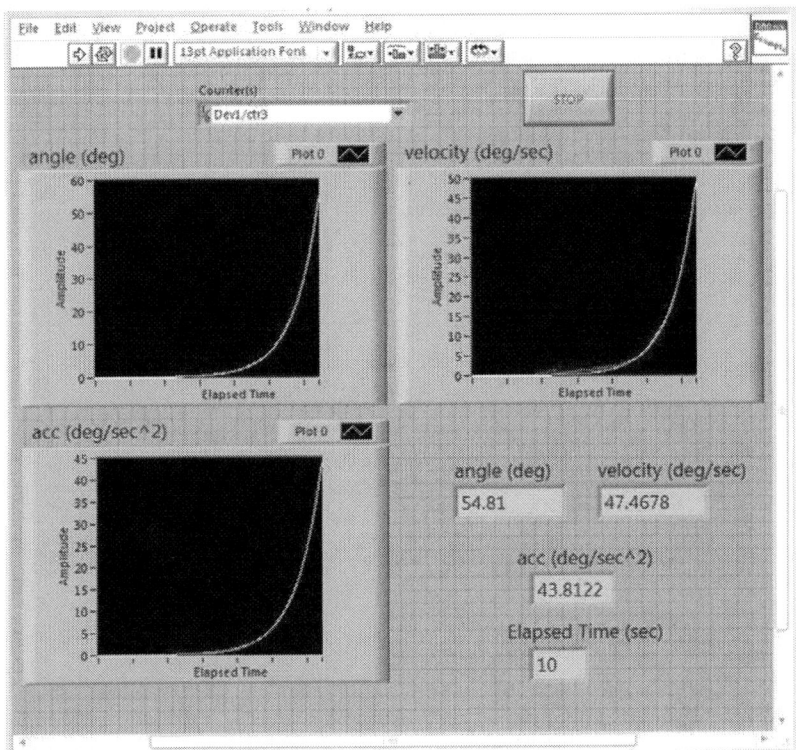

Figure 19. The values of the estimated and measured angular position, velocity and acceleration of the third joint angle (white: estimated, red: measured).

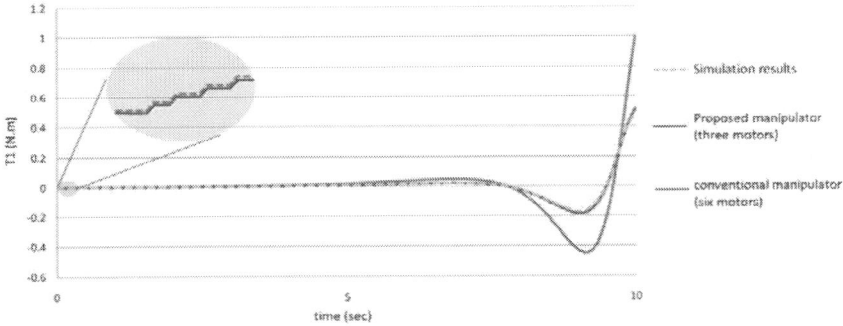

Figure 20. The torque of the first joint angle.

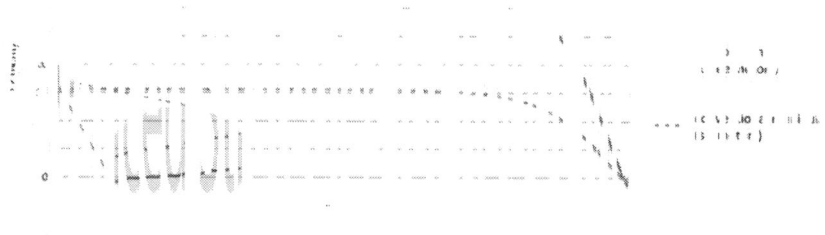

Figure 21. The torque of the second joint angle.

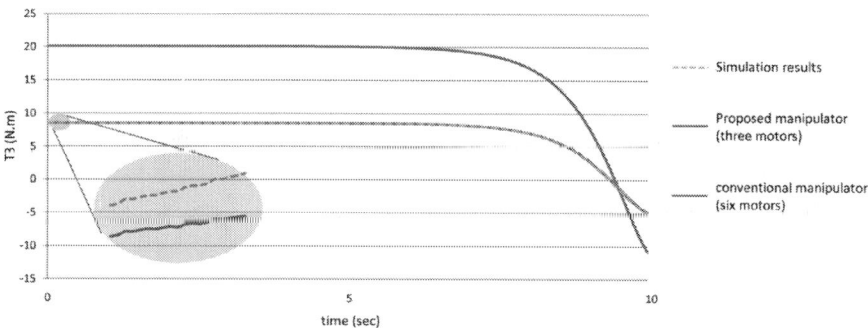

Figure 22. The torque of the third joint angle.

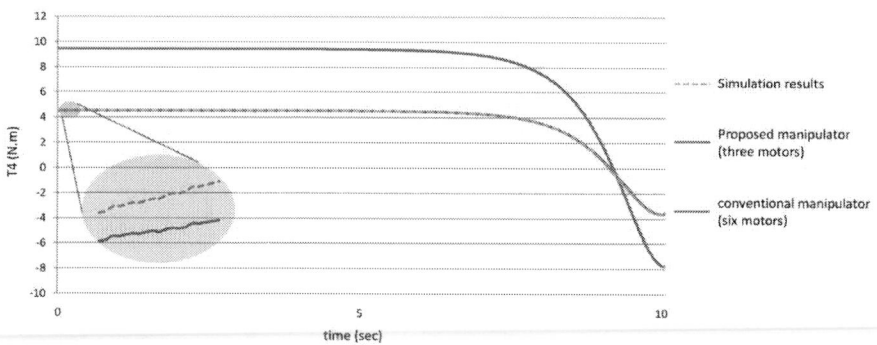

Figure 23. The torque of the fourth joint angle.

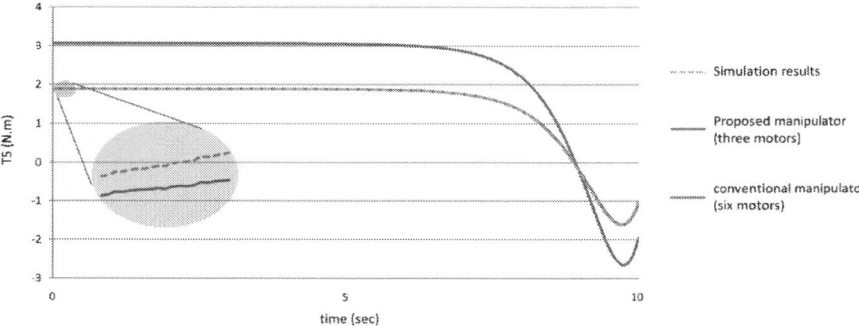

Figure 24. The torque of the fifth joint angle.

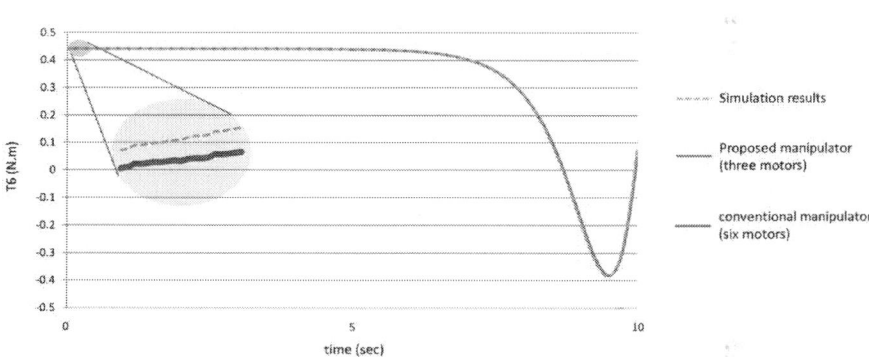

Figure 25. The torque of the sixth joint angle.

The results obtained from verification experiments indicate that there is a good agreement between the torque of the joint angles for the proposed manipulator using the estimated joint angles path (green) and the measured joint angles path (red). These figures show the effectiveness of the proposed manipulator in decreasing the torque of the joint angles using the proposed manipulator.

As mentioned in the first example that for the proposed manipulator, the third motor should balance the torque of all the third, fourth, fifth and the sixth joint, i.e., the torque of the third motor should equal $(T_3 + T_4 + T_5 + T_6)$ for the proposed manipulator, Figure 26 shows that even though that this motor (third motor) should balance the torques of four links, this motor could be smaller in size (less power) in the proposed manipulator than the third motor in the conventional three dimensional planar manipulator.

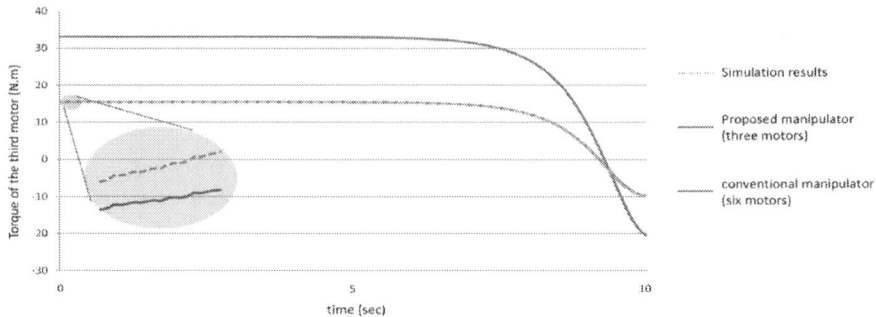

Figure 26. The torque of the third motor.

5. CONCLUSIONS

This paper presents a mechanical design for a three dimensional planar redundant manipulator. Theoretically, for each degree of freedom there should be one motor. However, in this design only three motors are needed to control any n degrees of freedom three dimensional planar redundant manipulator. Therefore, this design can be used to decrease the weight of the manipulator significantly. The design steps of this manipulator are explained in detail. The dynamical equations are calculated for both the proposed and the conventional three dimensional planar manipulators (with n motors) and it is concluded from the result, that even though the proposed manipulator has less motors, these motors could be even smaller (as regard to power) than the motors used with conventional three dimensional planar manipulators.

REFERENCES

1. Tsai, L.-W. Robot Analysis: The Mechanics of Serial and Parallel Manipulators; John Wiley & Sons, Inc.: Hoboken, NJ, USA, 1999.

2. Ma, S.; Yoshinada, H.; Hirose, S. CT ARM-I: Coupled tendon-driven manipulator model I-design and basic experiments. Proceeding of IEEE

International Conference on Robotics and Automation, Nice, France, 12–14 May 1992; Volume 3. pp. 2094–2100.

3. Ma, S.; Hirose, S.; Yoshinada, H. Design and experiments for a coupled tendon-driven manipulator. IEEE Control Syst. Mag. **1993**, 13, 30–36.

4. Feliu, V.; Ramos, F. Strain gauge based control of single-link flexible very lightweight robots robust to payload changes. Mechatronics **2005**, 15, 547–571.

5. Albu-Schaffer, A.; Haddadin, S.; Ott, Ch.; Stemmer, A.; Wimböck, T.; Hirzinger, G. The DLR lightweight robot: Design and control concepts for robots in human environments. Ind. Robot Int. J. **2007**, 34, 376–385.

6. Hagn, U.; Nickl, M.; Jorg, S.; Passig, G.; Bahls, T.; Nothhelfer, A.; Hacker, F.; Le-Tien, L.; Albu-Schaffer, A.; Konietschke, R.; et al. The DLR MIRO: A versatile lightweight robot for surgical applications. Ind. Robot: Int. J. **2008**, 35, 324–336.

7. Hirose, S.; Ma, S. Moray drive for multijoint. Proceeding of 5th International Conference of Advanced Robotics, Pisa, Italy, 19–22 June 1991; Volume 1. pp. 521–526.

8. Merlet, J.P. Parallel Robots, 2nd ed.; Springer: Dordrecht, The Netherlands, 2006.

9. Londi, F.; Pennestri, E.; Valentini, P.P.; Vita, L. Control and virtual reality simulation of tendon driven mechanisms. Multibody Syst. Dyn. **2004**, 12, 133–145.

10. Rodriguez-Donate, C.; Osornio-Rios, R.A.; Rivera-Guillen, J.R.; de Jesus Romero-Troncoso, R. Fused smart sensor network for multi-axis forward kinematics estimation in industrial robots. Sensors **2011**, 11, 4335–4357.

11. Lee, J.K.; Kim, K.; Lee, Y.; Jeong, T. Simultaneous intrinsic and extrinsic parameter identification of a hand-mounted laser-vision sensor. Sensors **2011**, 11, 8751–8768.

12. Yahya, S.; Moghavvemi, M.; Mohamed, H.A.F. Geometrical approach of planar hyper-redundant manipulators: Inverse kinematics, path planning and workspace. Simul. Model. Pract. Theory **2011**, 19, 406–422.

13. Mohamed, H.A.F.; Yahya, S.; Moghavvemi, M.; Yang, S.S. A new inverse kinematics method for three dimensional redundant manipulators. ICROS-SICE, Fukuoka, Japan, 18–21 August 2009; pp. 1557–1562.

14. Yahya, S.; Moghavvemi, M.; Mohamed, H.A.F. Singularity avoidance of a six degrees of freedom three dimensional redundant planar manipulator. Comput. Math. Appl. **2012**.

15. Yahya, S.; Moghavvemi, M.; Mohamed, H.A.F. A review of Singularity avoidance in the inverse kinematics of redundant robot manipulators. Int. Rev. Autom. Control **2011**, 4, 807–814.

16. Yahya, S.; Moghavvemi, M.; Mohamed, H.A.F. Redundant manipulators kinematics inversion. Sci. Res. Essays **2011**, 6, 5462–5470.

17. Khalil, W.; Dombre, E. Modeling, Identification & Control of Robots; Hermes Penton Ltd.: London, UK, 2002.

18. Siciliano, B.; Sciavicco, L.; Villani, L.; Oriolo, G. Robotics: Modelling, Planning and Control; Springer-Verlag London Limited: London, UK, 2009.

19. Siciliano, B.; Khatib, O. Springer Handbook of Robotics; Springer-Verlag: Berlin/Heidelberg, Germany, 2008.

20. Angeles, J. Fundamentals of Robotic Mechanical Systems: Theory, Methods, and Algorithms, 3rd ed.; Springer: Berlin/Heidelberg, Germany, 2007.

21. Spong, M.W.; Hutchinson, S.; Vidyasagar, M. Robot Dynamics and Control, 2nd ed.; John Wiley & Sons, Inc.: Hoboken, NJ, USA, 2004.

22. Shang, W.W.; Cong, S. Nonlinear computed torque control for a high speed planar parallel manipulator. Mechatronics **2009**, 19, 987–992.

23. Yoshida, K.; Kurazume, R.; Umetani, Y. Torque optimization control in space robots with a redundant arm. Proceeding of IEEE/RSJ International Workshop on Intelligent Robots and Systems, Osaka, Japan, 3–5 November 1991; Volume: 3. pp. 1647–1652.

CHAPTER 5

Nano-Workbench: A Combined Hollow AFM Cantilever and Robotic Manipulator

Héctor Hugo Pérez Garza [1,‡], Murali Krishna Ghatkesar [1,‡,*],
Shibabrata Basak [2], Per Löthman [3] and Urs Staufer [1]

[1] *Precision and Microsystems Engineering, Delft University of Technology,*
Mekelweg 2, 2628 CD Delft, The Netherlands
[2] *National Centre for HREM, Kavli Institute of Nanoscience Delft, Delft*
University of Technology, Lorentzweg 1, 2628 CJ Delft, The Netherlands
[3] *Korea Institute of Science and Technology, KIST Europe, Nano Magnetics*
group, Campus E7 1, 66123 Saarbrücken, Germany

ABSTRACT

To manipulate liquid matter at the nanometer scale, we have developed a robotic assembly equipped with a hollow atomic force microscope (AFM) cantilever that can handle femtolitre volumes of liquid. The assembly consists of four independent robots, each sugar cube sized with four degrees of freedom. All robots are placed on a single platform around the sample forming a nano-workbench (NWB). Each robot can travel the entire platform and has a minimum position resolution of 5 nm both in-plane and out-of-plane. The cantilever chip was glued to the robotic arm. Dispensing was done by the capillarity between the substrate and the cantilever tip, and was monitored visually through a microscope. To evaluate the performance of the NWB, we have performed three experiments: clamping of graphene with epoxy, mixing of femtolitre volume droplets to synthesize gold nanoparticles and accurately dispense electrolyte liquid for a nanobattery.

KEYWORDS

nano-workbench;   femtopipette;   nanorobot;   nanomanipulation; microfluidics; nanodispensing; hollow cantilever; atomic force microscope

1. INTRODUCTION

One of the goals within the field of nanotechnology is to develop novel tool systems and devices that are capable of manipulating objects at nanoscale with extreme precision and resolution [1]. In order to satisfy such critical requirement, different tools have been developed, among which the atomic force microscope (AFM) started to attract widespread attention due to numerous applications in biological and materials science technologies [2,3]. Soon after the invention of AFM, its further capabilities, beyond imaging, were realized. Consequently, researchers developed AFM-based nanomanipulation into a major technique that can be used in variety of applications [4,5]. However, the conventional AFM cantilevers have one sharp tip as the end-effector that can apply or detect a point force at the atomic scale. We have extended its capability from manipulating objects to manipulating small volume liquids. We have developed a hollow AFM cantilever that can handle femtolitre volumes of liquid [6,7]. It is not only used to interact with objects and surfaces by mechanical contact manipulation, imaging, force spectroscopy, but also to explore the possibilities of precise functionalization, dispensing/aspirating of different reagents and local/controlled mixing of droplets among many possibilities [8,9,10]. On the other front, robotics-based nanomanipulation is emerging as a new field that comprises design, modeling and fabrication of different nanotools to be used by robots, hence obtaining a programmable manipulation of matter at the nanoscale [11,12]. In the present work, we have combined both of these worlds to obtain robotic based precise nanomanipulation of femtolitre volume liquids. This combination enabled us to dispense liquids at unprecedented levels. We have attached our hollow AFM cantilever to an ultrasmall size piezoactuated sugar cube sized robot with a very small footprint. By combining multiple of these robots, we have made a robotic arena for micro and nanomanipulation that is referred to as the nano-workbench (NWB). The capability of the NWB is demonstrated by three experiments: clamping graphene, mixing femtolitre volume droplets to synthesize nanoparticles and precise placement of an electrolyte for nanobattery applications.

2. EXPERIMENTAL SECTION

2.1. Nano-Workbench (NWB)

In order to achieve high precision and accuracy during manipulation, we used the commercially available robots (miBot) from Imina Technologies (Figure 1a). Such robot is a mobile nanomanipulator with 4 degrees of freedom (DoF), two translations and two rotations. It combines long traveling range displacements with nanometer resolution. This motion is obtained by using the piezo-actuators placed on the bottom-side of the robot (Figure 1b). These actuators can be operated in stepping mode or scanning mode. In the stepping mode, a stick-slip method is used to move the robots for large distances obtained with a series of small strokes each in the range of several hundreds of nanometers. In the scanning mode, a high-resolution motion of few nanometers within one step is obtained. The robot has a piezo-electric actuated arm that moves in the vertical direction. In order to equip the robot to perform fluid manipulations, the previously developed hollow AFM cantilever was mounted on the arm (Figure 1c). To make these devices, two silicon substrates with reservoir defined in one and hollow portion of the cantilever in the other were fused. Later, silicon dioxide layer was grown inside the hollow channels. Then, the cantilevers were released to form a transparent hollow AFM cantilever with a silicon nitride tip [6]. The device can be used to dispense liquids in the sub-femtolitre regime and image surfaces with the tip. As shown in Figure 1c, our device resembles a commercial AFM-chip, but with an inlet at the on-chip fluid reservoir located on one side, which is connected to an outlet located at the tip via the hollow cantilever on the other side. When the inlet is filled, the liquid travels through the integrated capillaries of the cantilever until the aperture at the AFM tip. The liquid was dispensed by capillarity obtained by bringing the cantilever in contact with the substrate. The interplay between the substrate surface energy and the surface tension of the liquid controlled the amount of liquid that was dispensed. Typically, the dispensing can be monitored from the force-distance curves in an AFM system, however, in our case, we do not have any in situ detection mechanism to monitor dispensing. The dispensing is monitored through a microscope by looking at the snapping of the cantilever tip to the substrate due to capillarity when it is taken very close to the substrate. Then, lifting the robot arm retracted cantilever leaving a droplet on the substrate. The chip was epoxy glued to the robotic arm in such way that the cantilever tip would always point downwards, such that the aperture located in the tip could be brought in touch with the intended sample/surface to manipulate. No additional external fluid connections were made to the chip.

A control pad, which was plugged into a USB slot of the computer, allowed remote control of the nanomanipulators by increasing/decreasing the frequency and amplitude of the actuators for the vertical and horizontal motions. The external electronics, also plugged to the computer via a USB cable, was used for the control of piezoelectric actuators via flex cables. The robotic arena over which the robots were operating was made of a steel plate. An overview of the setup is shown in Figure 2. To monitor the performance of both the robot and hollow AFM cantilever during manipulation, the robotic arena is mounted either under an upright microscope with long working distance objectives or on top of an inverted fluorescence microscope to view through a transparent glass substrate and an additional top view camera. The speed and the resolution of positioning of the robot could be tuned by varying the frequency and the amplitude of the driving signals.

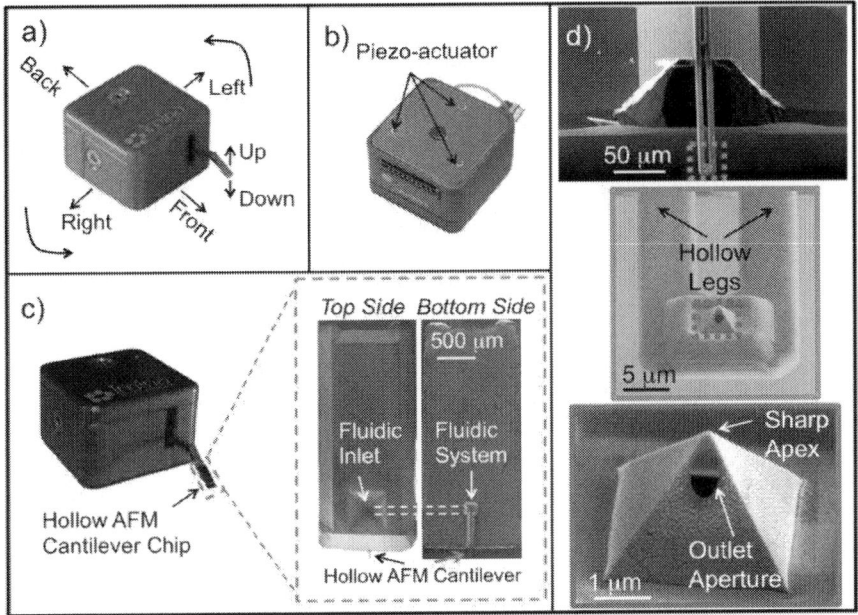

Figure 1. Combined Hollow AFM cantilever and Robotic Nanomanipulator System. (**a**) The robot has four degrees of freedom that gives it the versatility to move and/or position the mounted chip in the required place. (**b**) The bottom side of the robot has three piezo actuators that can be operated to give the desired motion. (**c**) The chip was mounted on the arm in the front to obtain the desired out of plane motion. All the motions can be remotely controlled. The fluid reservoir is located on the topside and the cantilever is on the bottom-side. A desired liquid can be loaded in the fluid reservoir. (**d**) The hollow cantilever is connected to the on-chip reservoir through the microfluidic channels. In a close view of the hollow cantilever, an aperture near the tip is shown. The aperture is made using a focused ion beam on the side of the pyramid to preserve the sharp apex for imaging.

2.2. Experiments with the Nano-Workbench

To demonstrate the capability and accuracy of using the NWB, we have performed three different experiments.

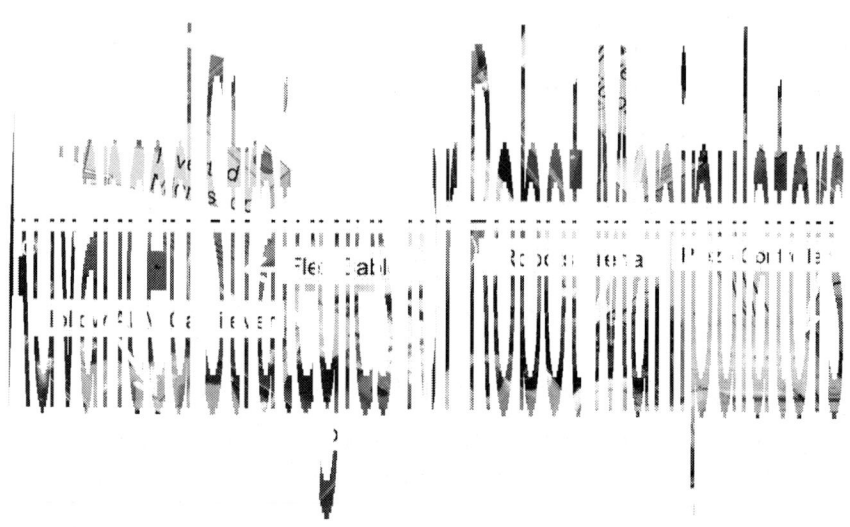

Figure 2. Nano-workbench. (**a**) Schematic of the working platform mounted on an inverted microscope with top view camera. (**b**) The picture of the setup. (**c**) Nanorobotic arena with four robots that can be operated in a time-multiplexed fashion. Robots are connected through a flex-cable to a hub, which in turn is connected to the computer. (**d**) The piezo controller and console for manipulation is shown.

1. **Clamping suspended graphene:** In a micro-electro-mechanical system (MEMS) based tensile tester, there are two suspended shuttle beams with a separation of about 5 microns. The poly-silicon chevron-type thermal actuators connected to one of the shuttle pulled the sample suspended between the shuttle beams. Care was taken to keep the sample at room temperature with heat sinks attached to the shuttle. The detailed design and fabrication of the device is given elsewhere [6]. The sample under investigation for their tensile properties is placed between the shuttles to bridge the separation. In our experiment, we have transferred graphene between the shuttles. The van der Waals forces with the shuttle surface held the graphene. The shuttles were actuated such that the graphene would experience a tensile force. However, the

graphene was always slipping away as the tensile force was overcoming the existing adhesion due to van der Walls forces between graphene and shuttle surface. To overcome this problem, clamping graphene was necessary. Using our NWB, we have locally dispensed epoxy on the shuttle-graphene interface present on both the shuttles of the MEMS device. The hollow cantilever was carefully brought to the point of interest by means of the remotely controlled robot. Once positioned in the vicinity of the graphene, the robot arm displaced the hollow cantilever slowly and accurately towards the edge of the graphene. Then, the hollow cantilever was snapped-in multiple times to locally dispense the epoxy along the edge of the shuttle over an area ranging about 10 µm × 10 µm (Figure 3). This area covered the entire graphene edge on the shuttle. The process was repeated on both edges, thus clamping graphene on both shuttles with epoxy. The graphene was stretched while recording the Raman spectra of the graphene. The 2D band and G band peaks were monitored in situ during stretching.

2. **Mixing microdroplets:** In this experiment, we have used the NWB to enable local deposition and controlled synthesis of gold nanoparticles. A chip containing a transmission electron microscopy (TEM) window of 50 nm thick Si_3N_4 and two different robots that were equipped with their own respective hollow cantilevers were used. One of them was filled with aqueous metallic salt solution (chloroauric acid–$AuCl_4$), while the other was filled with aqueous reducing agent solution (sodium azide–NaN_3). We first brought the robot carrying the metal salt solution near the TEM chip and approached the hollow cantilever towards the membrane window to dispense a droplet of controlled volume. After deposition, we brought the second robot containing the reducing agent and a droplet with the same volume was dispensed, allowing both reagents to coalesce and mix (Figure 4). Nanoparticles were synthesized in different droplet volumes ranging from 2.4 to 0.25 fL in equal quantities of both reagents on different windows. The results of the triggered chemical mixture were analyzed under TEM.

3. **Dispensing electrolyte for nanobattery:** In situ transmission electron microscope (TEM) with single grain electrode material can be used to visualize the lithium (de)intercalation process in real-time to understand the exact lithium transport mechanism inside the electrode material and thus better nanostructured electrode materials can be synthesized to improve the state of art lithium ion battery technology [6,13,14]. However, to prepare such a nanobattery assembly with single grain material, one needs to dispense tiny liquid electrolyte between the two electrodes with high accuracy [15]. The NWB was used to dispense not only tiny droplets of the ionic liquid (having low vapor pressure, stable in the high vacuum of TEM), namely 1 M lithium bis(trifluoromethane sulfonyl)imide (LiTFSI) in 1-Butyl-1-methylpyrrolidinium bis(trifluoromethylsulfonyl)imide, electrolyte but also precisely at the required position to complete the nanobattery assembly. Since the electrolyte was sensitive to air, dispensing was carried out inside the glove box, by placing the entire NWB setup inside the glove box. Battery assembly was made on a 20 nm silicon nitride membrane with appropriate electrical connections already fabricated before dispensing the electrolyte droplet.

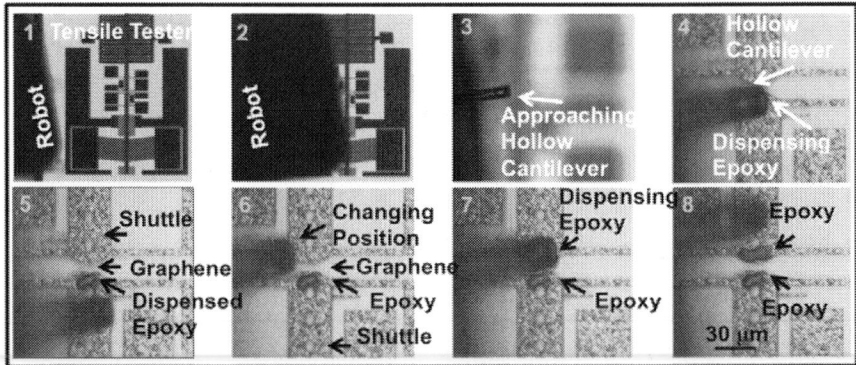

Figure 3. Clamping of graphene on a MEMS device. The optical microscope images (**1**) to (**8**) are the snapshots of events happening during the process of clamping graphene with an epoxy. (**1**) The overview of MEMS tensile tester. By thermal expansion, one shuttle pulls the sample and the sample bridging

the other shuttle moves. On-chip capacitive comb sensors detect motion of the shuttle. (**2**) Approaching with hollow cantilever to the location of dispensing. (**3**) Zoom into the approaching cantilever. (**4**) Dispensing epoxy on one of the shuttles. (**5**) Dispensed epoxy on one of the shuttle with graphene suspended between the shuttles. (**6**) Switching to the other shuttle. (**7**) Dispensing epoxy on the other shuttle. (**8**) The deposited epoxy on both shuttles, which eventually was cured and clamped the suspended graphene. The hollow cantilever attached to the robot arm was remotely controlled. The magnification for images (**3**) to (**8**) is same and the scale bar is given in (**8**).

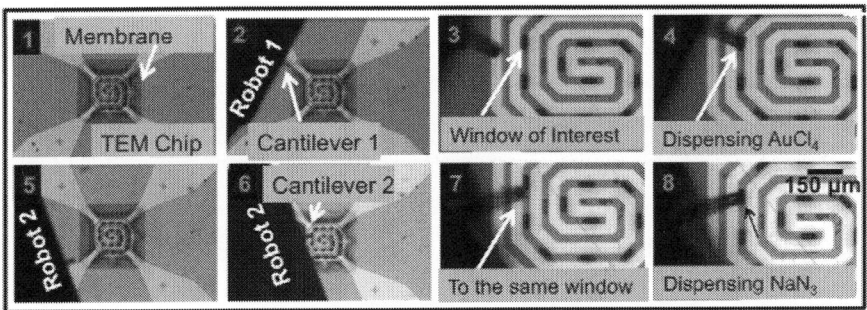

Figure 4. Dispensing to mix two droplets of reagents to synthesis gold nanoparticles. The optical microscope images (**1**) to (**8**) are the snapshots of the events in time during dispensing of droplets on a 50 nm thin silicon nitride window. Scale bar for panels (**3**), (**4**), (**7**) and (**8**) is shown in panel (**8**). See supplementary material for the video.

3. RESULTS AND DISCUSSION

The assembly of the NWB resulted in a well-mounted robotic arena on top of the inverted microscope that had an opened window in the center, such that the lens from the inverted microscope could have access to view the sample during manipulation. The robots were also portable enough to place them under upright microscope to visualize from top. The robots travel in two modes. On our setup, in the stepping mode, they made large distance motions covering a range of X: 25 mm, Y: 25 mm and Z: 8 mm, whereas in scanning mode they can be moved in the range of X: 150 nm, Y: 150 nm and Z: 300 nm. As for the positioning resolution, in stepping mode it was found to be X: 50 nm, Y: 50 nm and Z: 150 nm. Similarly, in scanning mode resolution was found to be X: 5 nm, Y: 5 nm, Z: 5 nm. The controller of the piezoactuators gave an output frequency of 0.7 Hz to 23.5 kHz and an output voltage ranging between 55–185 V (AC) and 0–185 V (DC).

The robots' ability to position in a given spot and move the arm up and down was due to the piezoelectric actuators located under the robot. Basically, the actuators rely on inertia and friction, in a mode referred to as the stick-and-slip mode. Due to the typically limited displacement of the actuators, stick-slip offered the possibility to combine very high resolutions with very large range of displacements. Therefore, if the robot needed to be driven over large distances, the piezo-stick-slip actuator generated a series of small steps, each with hundreds of nanometers in displacement. This is also referred to as the stepping mode. When high resolution was needed, the robot could be positioned within one step with a resolution of a few nanometers depending on the input control used in the scanning mode.

With very low vapor pressure and high viscosity of the epoxy, dip and dispense technique was used for these experiments. The successful clamping of the graphene using the developed NWB, allowed achieving extreme strains in graphene, which were one order of magnitude higher than the previously reported values. Before clamping, a maximum strain of ~1.3% was obtained, after clamping we could achieve strains of ~12.5% (Figure 5). A significant decrease in the amplitude of the G-band and red shift in the 2D band are the signatures of the stretched graphene [16,17]. A detailed analysis of the above behavior is reported elsewhere [18]. Furthermore, tensile testing experiments were also performed on samples with various number of graphene layers suspended between the shuttles and clamping by epoxy. An unprecedented strains of 14% for monolayers and 11% for four layer graphene were obtained respectively and reported elsewhere [19].

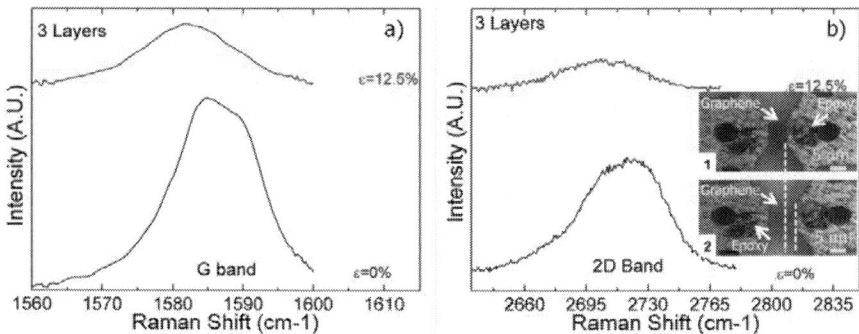

Figure 5. Raman spectra of graphene showing signatures of 3 layers of graphene at G band and 2D band. The inset picture in the right panel is the image of the shuttles with graphene suspended between them before (**1**) and after stretching (**2**). The dashed lines are drawn as a guide to the reader indicating the displacement obtained with stretching resulting in a strain of 12.5% in graphene in the maximum stretched condition. (**a**) A shift in the G band with significant decrease in amplitude confirms the stretched graphene. (**b**) Similarly, a red shift in the 2D band also confirms the stretching of the graphene.

The experiments to synthesis gold nanoparticles on a thin silicon nitride membrane were performed in a non-condensing humid environment. The dispensed droplets being very small evaporated in less than a second without the humid environment. The volume of the dispensed droplet was separately measured using force distance curves on an AFM setup and reported elsewhere [20]. The local dispensing of the metal salt solution and reducing agent with equal volumes resulted in controlled synthesis of gold nanoparticles (Figure 6). Different droplet sizes were dispensed resulting in different sizes of synthesized nanoparticles. The dispensing of the droplet was by capillarity. The amount of volume dispensed depends on the surface energy of the substrate, surface tension of the liquid and the contact time during dispensing. A detailed analysis of the particle synthesis in such small droplets is reported elsewhere [20].

We could place the electrolyte drop with sub micron accuracy to prepare nanobattery for TEM experiments (Figure 7). During charging of the battery, Li^+ ion will move out of the cathode, travel through the electrolyte and get deposited on the gold. While discharging, the Li^+ will travel back to the cathode. Hence, the gold line pattern on the topside (see Figure 7) also acts as a pseudo anode.

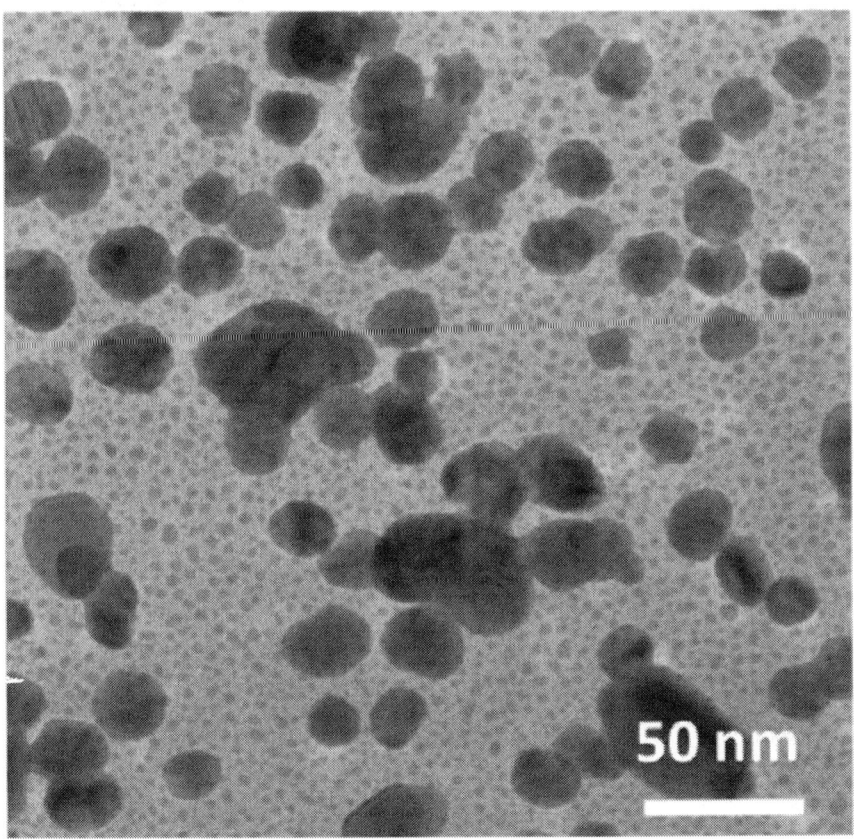

Figure 6. TEM image of the synthesized nanoparticles obtained by mixing femtolitre volume droplets of chloroauric acid and sodium azide. The image was taken with particles on a 50 nm silicon nitride membrane.

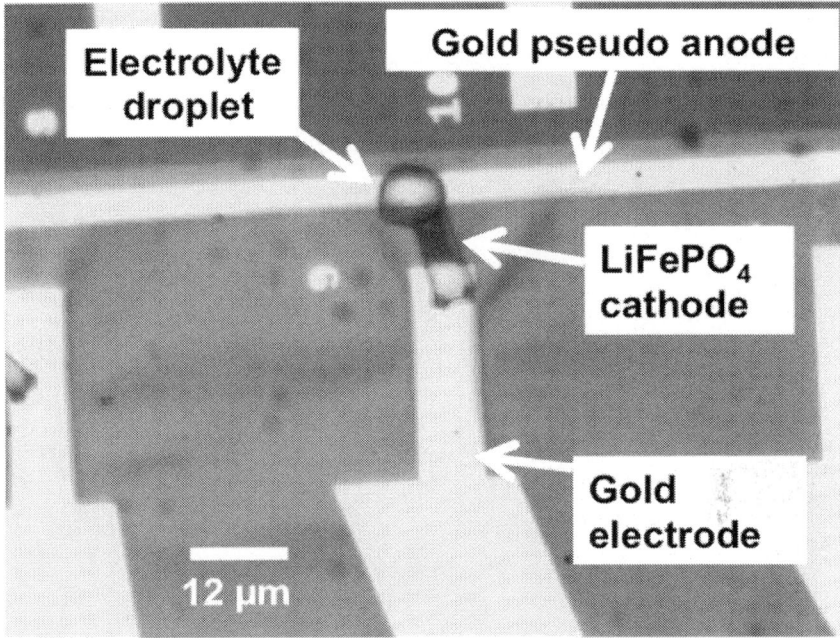

Figure 7. Electrolyte droplet of ~6 μm dispensed on a 20 nm silicon nitride membrane to connect a lamella of 12 μm × 4 μm × 100 nm LiFePO₄ cathode and gold pseudo anode electrode. The droplet is accurately dispensed between anode and cathode bridging a gap of 2 microns.

4. CONCLUSIONS

In conclusion, the hollow AFM cantilever mounted on a sugar cube sized robot can be used to locally and accurately dispense femtolitre volume liquid material on the target sites. A summary with the technical specifications of our NWB is shown in Table 1. Three experiments: Clamping of graphene on a MEMS tensile tester, synthesis of gold nanoparticles in femtolitre volume droplets on a 50 nm thin silicon nitride membrane and accurate dispensing of electrolyte droplets on a 20 nm thin silicon nitride membrane were performed to show the capabilities of the NWB. A significant improvement from 2% to 12.5% strain in three-layer graphene was obtained after clamping. As an outlook, the NWB provides an opportunity to study the mechanical properties (buckling, adhesion) of different samples (i.e., carbon nanotubes, graphene) and also to evaluate the behavior of a biological cell as a function of mechanical/chemical stimuli. By integrating piezoresistors on the attached hollow AFM cantilever [21], the amount of interface forces involved during fluid-fluid and fluid-surface interactions could be quantified. Furthermore, with on chip fluid reservoir interfaced to the external pressure control system, dispensing and aspiration can be well controlled. Other examples of manipulation of matter at this particular scale might include functionalization and assembly of micro and nano-electromechanical systems.

Table 1. Summary of technical specifications

Hollow AFM Cantilever
Resonance Frequency: 153.94 kHz
Spring Constant: 3.48 N/m
Typical Dispensed Volume: 30 fL
Storage Volume: 20 nL
Aperture Diameter: 300–2000 nm
Robot
Travelling Range
(a) Stepping Mode: X: 25 mm, Y: 25 mm, Z: 8 mm
(b) Scanning Mode: X: 150 nm, Y:150 nm, Z: 300 nm
Positioning Resolution
(a) Stepping Mode: X: 50 nm, Y: 50 nm, Z: 150 nm
(b) Scanning Mode: X: 5 nm, Y: 5 nm, Z: 5 nm
Controller
Output Frequency: 0.7 Hz to 23.5 kHz
Output Voltage: 55–185 V (AC) and 0–185 V (DC)

ACKNOWLEDGMENTS

This work is supported by NanoNextNL, a micro and nanotechnology consortium of the government of the Netherlands and 130 partners.

AUTHOR CONTRIBUTIONS

Héctor Hugo Pérez Garza, Murali Krishna Ghatkesar, Shibabrata Basak performed the experiments. Per Löthman contributed to gold nanoparticle synthesis chemistry. Urs Staufer conceived the idea of nano-workbench. Héctor Hugo Pérez Garza and Murali Krishna Ghatkesar wrote the manuscript. All authors commented and agreed with the content.

REFERENCES

1. Research, I.A. Boy and His Atom: The World's Smallest Movie. Available online: Http://www.research.ibm.com/articles/ madewith atoms.shtml (accessed on 27 February 2015).

2. Müller, D.J.; Dufrêne, Y.F. Atomic force microscopy: A nanoscopic window on the cell surface. Trends Cell. Biol. **2011**, 21, 461–469.

3. Mohn, F.; Gross, L.; Moll, N.; Meyer, G. Imaging the charge distribution within a single molecule. Nat. Nanotechnol. **2012**, 7, 227–231.

4. Li, G.; Xi, N. Overview of Nanomanipulation by Scanning Probe. In Introduction to Nanorobotic Manipulation and Assembly; Artech House: Norwood, MA, USA, 2012; pp. 93–110.

5. Xi, N.; Fung, C.; Yang, R.; Seiffert-Sinha, K.; Lai, K.W.C.; Sinha, A.A. Bionanomanipulation using atomic force microscopy. IEEE Nanotechnol. Mag. **2010**, 4, 9–12.

6. Ghatkesar, M.K.; Garza, H.H.P.; Staufer, U. Hollow AFM cantilever pipette. Microelectron. Eng. **2014**, 124, 22–25.

7. Ghatkesar, M.K.; Garza, H.H.P.; Heuck, F.; Staufer, U. Scanning Probe Microscope-Based Fluid Dispensing. Micromachines **2014**, 5, 954–1001.

8. Zambelli, T.; Meister, A.; Gabi, M.; Behr, P.; Studer, P.; Voros, J.; Niedermann, P.; Bitterli, J.; Polesel-Maris, J.; Liley, M.; Heinzelmann, H. Fluid FM: Combining Atomic Force Microscopy and Nanofluidics in a

Universal Liquid Delivery System for Single Cell Applications and Beyond. Nano Lett. **2009**, 9, 2501–2507.

9. Shibata, T.; Nakamura, K.; Horiike, S.; Nagai, M.; Kawashima, T.; Mineta, T.; Makino, E. Fabrication and characterization of bioprobe integrated with a hollow nanoneedle for novel AFM applications in cellular function analysis. Microelectron. Eng. **2013**, 111, 325–331.

10. Kang, W.M.; Yavari, F.; Minary-Jolandan, M.; Giraldo-Vela, J.P.; Safi, A.; McNaughton, R.L.; Parpoil, V.; Espinosa, H.D. Nanofountain Probe Electroporation (NFP-E) of Single Cells. Nano Lett. **2013**, 13, 2448–2457.

11. Xie, H.; Onal, C.; Régnier, S.; Sitti, M. Automated Control of AFM Based Nanomanipulation. In Atomic Force Microscopy Based Nanorobotics; Springer: Berlin, Germany, 2012; pp. 237–311.

12. Hou, J.; Liu, L.; Wang, Z.; Wang, Z.; Xi, N.; Wang, Y.; Wu, C.; Dong, Z.; Yuan, S. Afm-based robotic nano-hand for stable manipulation at nanoscale. IEEE Trans. Autom. Sci. Eng. **2013**, 10, 285–295.

13. Scrosati, B.; Garche, J. Lithium batteries: Status, prospects and future. J. Power Sources **2010**, 195, 2419–2430.

14. Liu, X.H.; Huang, J.Y. In situ TEM electrochemistry of anode materials in lithium ion batteries. Energy Environ. Sci. **2011**, 4, 3844–3860.

15. Wu, H.; Cui, Y. Designing nanostructured Si anodes for high energy lithium ion batteries. Nano Today **2012**, 7, 414–429.

16. Mohiuddin, T.; Lombardo, A.; Nair, R.; Bonetti, A.; Savini, G.; Jalil, R.; Bonini, N.; Basko, D.; Galiotis, C.; Marzari, N. Uniaxial strain in graphene by Raman spectroscopy: G peak splitting, Grüneisen parameters, and sample orientation. Phys. Rev. B **2009**, 79, 205433.

17. Ni, Z.H.; Yu, T.; Lu, Y.H.; Wang, Y.Y.; Feng, Y.P.; Shen, Z.X. Uniaxial strain on graphene: Raman spectroscopy study and band-gap opening. ACS Nano **2008**, 2, 2301–2305.

18. Pérez Garza, H.H.; Kievit, E.W.; Schneider, G.F.; Staufer, U. Controlled, Reversible, and Nondestructive Generation of Uniaxial Extreme Strains (>10%) in Graphene. Nano Lett. **2014**, 14, 4107–4113.

19. Pérez-Garza, H.H.; Kievit, E.W.; Schneider, G.F.; Staufer, U. Highly strained graphene samples of varying thickness and comparison of their behaviour. Nanotechnology **2014**, 25, 465708.

20. Garza, H.H.P.; Ghatkesar, M.K.; Löthman, P.; Manz, A.; Staufer, U. Enabling local deposition and controlled synthesis of Au-nanoparticles using a femtopipette. In Proceedings of the 2014 9th IEEE International Conference on Nano/Micro Engineered and Molecular Systems (NEMS), Waikiki Beach, HI, USA, 13–16 April 2014; pp. 323–328.

21. Garza, H.; Stoute, R.; Ghatkesar, M.; Staufer, U. Self-sensing nanopipette for liquid dispensing and AFM imaging. In Proceedings of the 2013 Transducers & Eurosensors XXVII: The 17th International Conference on Solid-State Sensors, Actuators and Microsystems (TRANSDUCERS & EUROSENSORS XXVII), Madrid, Spain, 16–20 June 2013; pp. 2648–2651.

CHAPTER 6

Experimental Investigation on Adaptive Robust Controller Designs Applied to Constrained Manipulators

Samuel L. Nogueira [1,*], **Tatiana F. P. A. T. Pazelli** [2], **Adriano A. G. Siqueira** [1] **and Marco H. Terra** [3]

[1] *Department of Mechanical Engineering, University of São Paulo, 400 Trabalhador São-Carlense av., 13566-590, São Carlos, Brazil*

[2] *Department of Electrical Engineering, Federal University of São Carlos, Washington Luís, km 235-SP-310, São Carlos, Brazil*

[3] *Department of Electrical Engineering, University of São Paulo, 400 Trabalhador São-Carlense av., 13566-590, São Carlos, Brazil*

ABSTRACT

In this paper, two interlaced studies are presented. The first is directed to the design and construction of a dynamic 3D force/moment sensor. The device is applied to provide a feedback signal of forces and moments exerted by the robotic end-effector. This development has become an alternative solution to the existing multi-axis load cell based on static force and moment sensors. The second one shows an experimental investigation on the performance of four different adaptive nonlinear H_∞ control methods applied to a constrained manipulator subject to uncertainties in the model and external disturbances. Coordinated position and force control is evaluated. Adaptive procedures are based on neural networks and fuzzy systems applied in two different modeling strategies. The first modeling strategy requires a well-known nominal model for the robot, so that the intelligent systems are applied only to estimate the effects of uncertainties, unmodeled dynamics and external disturbances. The second strategy considers that the robot model is completely unknown and, therefore, intelligent systems are used to estimate these dynamics. A comparative study is conducted based on experimental implementations performed with an actual planar manipulator and with the dynamic force sensor developed for this purpose.

KEYWORDS

constrained manipulators; $\mathcal{H}_\infty$ control; neural networks; fuzzy systems; variable structure control; load cell; multi-axis force sensor

1. INTRODUCTION

The first robotic manipulators were developed in order to perform positioning tasks. They were then specifically designed to be robust enough so as to not be affected by external disturbances. This physical robustness of robot manipulators has enabled researchers to obtain accurate positioning systems based on simple control laws. Decades later, the popularization of industrial robotics has heightened researchers' interest in creating a much wider range of applications for robotic manipulators in various environments.

Nowadays, many applications demand robotic manipulators to perform tasks subject to force and motion constraints. For example, the process of milling a piece requires accurate incidence angles, paths, forces and moments exerted by the drill in the milled material. Additionally, in industrial assembly lines, the objects must be assembled along certain paths with predetermined forces and moments. In sheet metal cutting, the cutting angles, paths and forces exerted on the material are also important.

Moreover, on surfaces where polishing disks must always be perpendicular to the surface being polished, predetermined force must be applied. Consequently, new concepts of position and force control for lighter and more flexible robots have been created [1,2].

The problem defined in these applications involves three stages: the approach phase, the impact moment and the sustained contact tracking. The approach phase has been addressed in many works and defines the problem of positioning the tool without, or before, touching the environment. The second phase requires controlling the initial impact and damping out the vibrations generated during the event. After the initial impact, sustained contact is desired in many operations. In these cases, not only the motion of the end-effector is required to follow a prescribed path, but also the force exerted by the end-effector is required to follow a predefined reference. In these constrained systems, forces and moments generated between the end-effector and the target must be controlled, rather than being treated as disturbances and rejected. Addressing manipulators subject to model uncertainties and disturbances, the work considered in this paper is concentrated on the sustained contact tracking phase of the problem.

The great majority of solutions presented in the literature for the problem of constrained systems control require knowing the forces and moments of interaction between the robot end-effector and the environment where it is acting. The problem of force control undertaken in this paper was also considered in [1,3–7]. However, only simulation results were presented.

As a differential and important contribution, this paper addresses an experimental investigation on robust force control as a result of the development of a modular sensor device. The proposed device is designed and built to measure dynamic forces and moments in three orthogonal axes based on unidirectional force sensor units. As a consequence of its independent architecture on the type of sensitive material, static or dynamic force sensors can be applied. Piezoelectric or piezoresistive force sensors are effective solutions for the applications involved in this study due to their inherent dynamic response characteristics.

This paper is organized as follows: next section presents preliminary concepts and relevant results found in the literature; Section 3 introduces the model description of the constrained robot manipulator; Section 4 presents the problem formulation; Section 5 describes the solutions for the nonlinear H_∞ control problems based on the linear parametrization property of the model, neural networks and fuzzy systems; Section 6 demonstrates the 3D dynamic force and moment sensor; and Section 7 presents the experimental results for a three-link manipulator.

2. PRELIMINARY CONCEPTS

The concept of stiffness control was introduced by Salisbury [8]. It is based on the resistance of the environment in which the robotic end-effector applies the force. The problem is modeled as a mass-spring system and this method made possible the simultaneous position/force control. However, it considers constant desired position and force. In many robotic applications, such as when milling a piece, the end-effector must follow a trajectory along the surface of an object while applying a desired force, which is not necessarily constant. In this case, the stiffness control application does not work properly.

To address this type of limitation, Raibert and Craig [9] partitioned the control problem into two subtasks: one task is for controlling the position trajectory and the other task for controlling the desired force. This approach has been evolutionary for controllers, as proposed by Paul et al. [10], and became the conceptual basis of the hybrid trajectory of position and force control currently found in the literature.

It was shown by McClamroch and Wang [11], that when a manipulator is in contact with a surface, the position degrees of freedom are reduced. In this case, force constraint is added to motion equations through Lagrange multipliers. Thus, the order of the state vector is reduced in the dynamic equations of the manipulator.

Since the phase of controlling robotic systems with restricted position and force has been overcome, researchers began to focus on variables that could degrade the stability of the proposed models. The load on the end-effector that can fluctuate while the manipulator performs various tasks, friction and parameter uncertainties, are some examples that have required much research effort [12–15]. However, only a few studies have dealt with adaptive and robust control of robotic systems subject to constraints [16–20]. Using the basis developed by [9,11], the work performed by Chang and Chen [21] submitted an adaptive controller with H_∞ robust performance criterion for robotic systems with position and force constraints.

On the development of adaptive robust controllers, neural networks, combined with a nonlinear H_∞ control law, were applied by Chang and Chen [22] to control robotic systems subject to parametric uncertainties and external disturbances. A smooth response was achieved with a simple and computationally efficient implementation. Neural networks were applied to estimate the unknown dynamics of the system, thereby not requiring mathematical modeling knowledge. A variable structure controller (VSC) is added to this formulation by Chang and Chen [23]. The inclusion of VSC in the control law weakens the hypothesis used by the authors in [22] that the estimation error should be integrable, and limits it to be only a state-dependent function. Following this approach, the authors in [24] developed an adaptive H_∞ controller based on fuzzy systems and VSC for robots with position and force constraints.

Karayiannidis and Doulgeri proposed adaptive controllers in [5,6] for force and position trajectory tracking in environments with little or no constraint knowledge. In these papers, the main purpose is to explore the workspace using measurements of position, speed and force in a robotic end-effector, to attenuate impacts caused by unknown environments. In order to improve estimates of these models, the use of a camera was proposed by Cheah et al. [4] to provide a better estimate of the contact constraint through computer vision. However, these works are more interested in exploring the environmental constraint, rather than a repetitive task control with great tracking precision.

To satisfy the need of measuring the forces and moments of interaction between the robot end-effector and the environment, devices known as multi-axis load cell or multi-axis force and moment sensor are used. There are several patents on devices whose purpose is to measure forces and moments in three axes.

Force sensors, such as strain gages, have been generally used as the basic unit of measurement, as shown by the illustrative examples in Figure 1. The multi-axis load cell developed by Meyer and Lowe [25] was built in one piece with internal and external parts that are connected by a pair of axially spaced beams. The beams are fixed in the center of the piece, where the strain gages are attached, and an outside tunnel. The loads are measured by the curvature of the connectors. The load transducer developed by Meyer et al. [26] measures linear forces in three axes and moments of about two axes. The transducer has encapsulations connected by internal and external arms-sensitive loading. The device described by Sommerfeld et al. [27] consists of an external annular body, a hub and four beams that hold the hub to the radially outward portion. Strain gages are fixed to the faces of the beams, with a 90 degree lagging. Forces and moments exerted on the hub are transmitted to the four beams, and consequently, to the strain gages. Differently from [25,26], this device described by Sommerfeld et al. [27] is able to measure forces and moments in three orthogonal axes. However, all the devices mentioned above have an extrusion completely dependent upon the type of sensitive material being used.

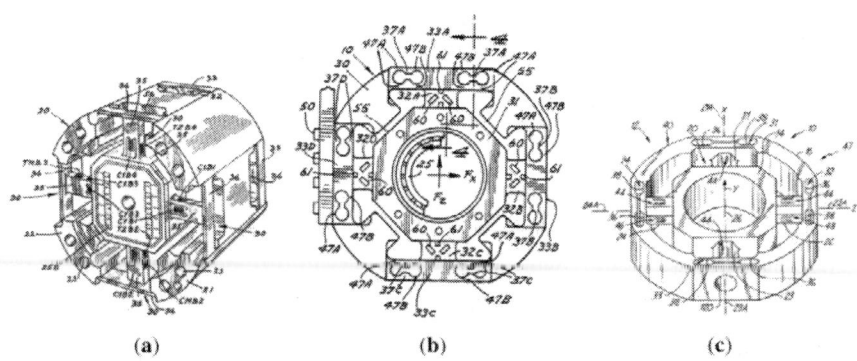

Figure 1. (**a**) Image taken from [25]; (**b**) Image taken from [26]; (**c**) Image taken from [27].

3. MODEL DESCRIPTION

Let a constrained robot manipulator be defined by an n-link serial-chain rigid manipulator whose end-effector is in contact with a rigid environmental constraint, according to Figure 2.

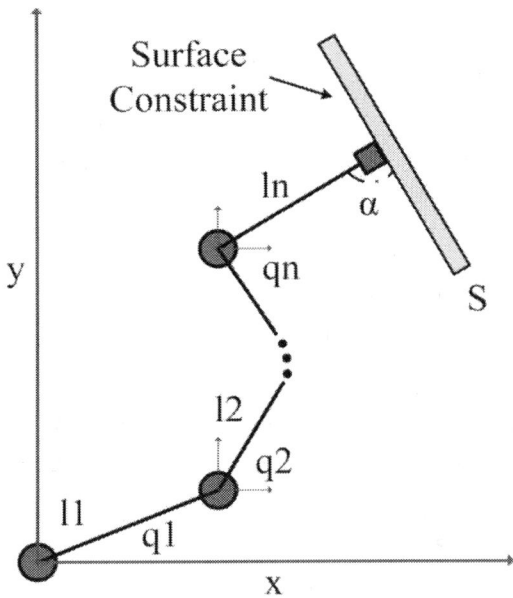

Figure 2. Constrained robot manipulator.

The links of the manipulator are numbered from 1 to n: J_i is the joint connecting the $(i - 1)$-th and i-th links, l_i is the vector connecting J_i and J_{i+1}, q_i is the angle of the i-th link around joint J_i, C_i is the center of mass of the i-th link and l_{ci} is the vector connecting J_i and C_i. The constraint surface is represented by S and α is the angle of contact between the end-effector and S.

Remark 1 *Assume that the end-effector is already in contact with the constraint surface, and the control exerted over the constraint force is such that the force will always maintain the end-effector in contact with the constraint surface.*

3.1. Robot Dynamics

The dynamic equations of a constrained robot is given from Lagrange theory as

$$M(q)\ddot{q} + C(q,\dot{q})\dot{q} + G(q) = \tau + f + \tau_d$$

(1)

where $M(q) \in \Re^{n \times n}$ is the symmetric positive definite inertia matrix, $C(q,\dot{q}) \in \Re^{n \times n}$ is the Coriolis and centripetal matrix, $G(q) \in \Re^n$ is the vector of the gravitational torques, $\tau \in \Re^n$ is the torque vector acting upon the manipulator joint, $f \in \Re^n$ denotes the vector of joint-space generalized forces on the environmental constraint exerted by the end-effector, and τ_d defines finite energy unknown disturbances.

Model uncertainties in Equation (1) can be introduced dividing the matrices M(q), C(q, q'), G(q), and f into a nominal and a perturbed part:

$$M(q) = M_0(q) + \Delta M(q)$$
$$C(q,\dot{q}) = C_0(q,\dot{q}) + \Delta C(q,\dot{q})$$
$$G(q) = G_0(q) + \Delta G(q)$$
$$f = f_0 + \Delta f$$

(2)

where $M_0(q), C_0(q,\dot{q}), G_0(q)$, and f_0 are nominal matrices and $\Delta M(q), \Delta C(q,\dot{q}), \Delta G(q)$, and Δf represent the uncertainties.

3.2. Constraint Modelling

Considering **Remark 1**, the m-dimensional surface constraint is described by the holonomic relationship

$$\phi(q) = 0$$

(3)

where $\phi(q) : \Re^n \to \Re^m$ is a smooth function.

Constraint forces are given by

$$f = J_c^T(q)\lambda$$

$$\tag{4}$$

where $J_c(q) = \frac{\delta\phi(q)}{\delta q} \in \Re^{m\times n}$ is the Jacobian matrix that relates the constraint to the controlled variables of the robot and $\lambda \in \Re^m$ is a vector of generalized Lagrangian multipliers associated with the constraint.

In this paper, it is considered that parametric uncertainties may also be included into the constraint model since the constraint surface may be not perfectly rigid, frictionless or even that its geometric description may not be exactly known. Thus, consider

$$\phi(q) = \Delta\phi(q)$$
$$J_c(q) = J_c(q) + \Delta J_c(q)$$

$$\tag{5}$$

and assume that $\Delta\phi$ and ΔJ_c are implicit in Δf, described in Equation (2).

3.3. Reduced Order Model

The presence of m constraints causes the manipulator to lose m degrees of freedom, and therefore, n – m linearly independent coordinates are sufficient to characterize the constrained movement. Therefore, to formulate a reduced order dynamics for the constrained system, the formulation presented in this paper follows the assumptions made by McClamroch and Wang [11] and posteriorly by Chang and Chen [24].

Define

$$q = \begin{bmatrix} q^1 \\ q^2 \end{bmatrix}$$

where $q^2 = \sigma(q^1)$ and $\phi(q^1, \sigma(q^1)) = 0$.

The following reduced model formulation is obtained for the constrained manipulator as the model proposed by Chang and Chen [24]:

$$A_L(q^1)\ddot{q}^1 + L^T(q^1)C_L(q^1, \dot{q}^1)\dot{q}^1 + L^T(q^1)G(q^1) = L^T(q^1)(\tau + \tau_d) \tag{6}$$

where

$$L(q^1) = \begin{bmatrix} I_{(n-m)} \\ \frac{\partial\sigma(q^1)}{\partial q^1} \end{bmatrix}$$

and

$$C_L(q^1, \dot{q}^1)\dot{q}^1 = M(q^1)\dot{L}(q^1) + C(q^1, \dot{q}^1)L(q^1)$$

The necessary structure and properties of the model for controller formulation are maintained since $A_L(q^1) = L^T(q^1)M(q^1)L(q^1)$ is symmetric positive definite and the matrix $\dot{A}_L(q^1) - 2L^T(q^1)C_L(q^1, \dot{q}^1)$ is skew symmetric.

4. PROBLEM FORMULATION

Let $q_d(t) \in \Re^n$ and $\dot{q}_d(t) \in \Re^n$ be the desired reference trajectory and the corresponding velocity for the joints, respectively Assume that $q_d(t)$ and its derivatives $\dot{q}_d(t)$ and $\ddot{q}_d(t)$ are bounded. Define a bounded $f_d \in \Re^n$ as the desired reference contact force. To be consistent with the imposed restrictions, we need to assure that $\phi(q_d) = 0$ and $f_d = J_c^T(q_d)\lambda_d$.

Since $q^2 = \sigma(q^1)$, it is only necessary to find a control law that makes $q^1 \to q_d^1$ when t → ∞. Therefore, define the position tracking error $\overline{x}_1(t)$ and the filtered link tracking error $\overline{x}_1(t)$ as in [24]:

$$\overline{x}(t) \doteq \begin{bmatrix} \overline{x}_1(t) \\ \overline{x}_2(t) \end{bmatrix} \doteq \begin{bmatrix} q^1(t) - q_d^1(t) \\ \dot{q}^1(t) - \dot{q}_d^1(t) + p(q^1(t) - q_d^1(t)) \end{bmatrix}$$

(7)

for some constant p > 0.

From Equations (6) and (7), the error dynamic equations can be obtained as

$$\dot{\overline{x}} = \begin{bmatrix} \dot{\overline{x}}_1 \\ \dot{\overline{x}}_2 \end{bmatrix} = A\overline{x} + Bu + B\omega$$

(8)

where

$$A = \begin{bmatrix} -pI & I \\ 0 & -A_L^{-1}(q^1)C_L(q^1, \dot{q}^1) \end{bmatrix}, \quad B = \begin{bmatrix} 0 \\ -(M(q^1)L(q^1))^{-1} \end{bmatrix}$$

$$u = F(x_e) - \tau, \quad \omega = \tau_d, \quad \text{and}$$

$$F(x_e) \doteq M(q^1)L(q^1)(\ddot{q}_d^1 - p\dot{\overline{x}}_1) + C_L(q^1, \dot{q}^1)(\dot{q}_d^1 - p\overline{x}_1) + G(q^1)$$

Within this problem formulation, the torques applied to the joints to guarantee the task execution are given by

$$\tau = F(x_e) - u \qquad (9)$$

where the term $F(x_e)$ refers to the dynamics of the controlled variables and u is the control law provided by the adaptive $\mathcal{H}_\infty$ controller proposed by Chang and Chen [24].

The term $F(x_e)$ can be divided in a nominal and uncertainty part, that is,

$$F(x_e) = F_0(x_e) + \Delta F(x_e)$$

where term $\Delta F(x_e)$ contains the uncertainties from parametric Equations (2) and (5).

In this paper, the control problem is solved based on two different approaches. In the first approach, an adaptive intelligent system is applied to estimate only the term $\Delta F(x_e)$, considering that the nominal model of the robot is well known. In the second approach, it is considered that the system model or the term $F_0(x_e) + \Delta F(x_e)$ is completely unknown, then the adaptive intelligent system is applied to estimate it. The nonlinear H$_\infty$ control and VSC are applied in both approaches to attenuate the effects of estimation errors and external disturbances.

5. ADAPTIVE NONLINEAR H$_\infty$ CONTROLLER BASED ON INTELLIGENT SYSTEMS

The adaptive control laws presented as follows are based on two different learning methods to estimate uncertain parameters and also the behavior of unmodeled dynamics. Neural networks and fuzzy systems based on the Takagi–Sugeno model are considered. The structure of these intelligent systems can be seen in details in [28].

Let $\hat{F}(x_e, \Theta) = \Xi(x_e)\Theta$ be the output of the adaptive intelligent system, where x_e is the input vector and Θ is a vector of adjustable parameters such that $\Theta^T\Theta \leq M_\theta | M_\theta > 0$. Two different cases are considered in the following:

Case 1 *Estimation of Model Uncertainties Based on Intelligent Systems*

In this case, the nominal model of the robot is considered as well known, the intelligent system then estimates only the model uncertainties, such as

$$\hat{F}_1(x_e, \Theta) \approx \Delta F(x_e)$$

Therefore u and ω in Equation (8) will be rewritten as

$$u = F_0(x_e) + \hat{F}_1(x_e, \Theta) - \tau$$
$$\omega = (\Delta F(x_e) - \hat{F}_1(x_e, \Theta)) + \tau_d$$

so that ω includes the estimation error from the intelligent system.

Case 2 *Estimation of Complete Model Based on Intelligent Systems*

In this case, however, the nominal model of the robot is considered completely unknown, so the intelligent system estimates the complete model, such as

$$\hat{F}_2(x_e, \Theta) \approx F_0(x_e) + \Delta F(x_e)$$

Therefore u and ω in Equation (8) is rewritten as

$$u = \hat{F}_2(x_e, \Theta) - \tau$$
$$\omega = (F(x_e) - \hat{F}_2(x_e, \Theta)) + \tau_d$$

so that ω includes the estimation error from the intelligent system.

Regarding the nonlinear $\mathcal{H}_\infty$ control solution proposed by Chang and Chen [24] for constrained systems, define $u = \bar{u}$ where $\bar{u} = u_P + u_F$, such that

$$u_P = k_0 E \bar{x}_2 - k(x_e) sgn(L\bar{x}_2) \tag{10}$$

$$u_F = J_c^T \lambda_c \tag{11}$$

where u_P is the $\mathcal{H}_\infty$+VSC control term for the position enforcement and u_F is the H$_\infty$ control law for the force tracking procedure, with

$$E := \begin{bmatrix} I_{(n-m)} \\ 0_{m\times(n-m)} \end{bmatrix} \quad \text{and} \quad \lambda_c \doteq \lambda_d - k_\lambda \int_0^T (\lambda - \lambda_d) dt$$

for some constant gain $k_0, k(x_e) > 0$ and $k_\lambda > 0$.

Theorem 1 presented in the following is a variation of the results presented in [21,23,24] with the difference that the nominal model and an intelligent system are considered rather than the linear parametrization of the robot. Thus, this theorem defines an adaptive H$_\infty$ controller to solve the same problem's position/force tracking control problems presented in these papers but with the advantage of using the known model of the robot.

Theorem 1 *Given a desired disturbance attenuation level $\gamma > 0$, a weighting matrix Q and the Lyapunov candidate function V(t), the following performance criterion*

$$\int_0^T \|\bar{x}(t)\|_Q^2 \leq V(0) + \gamma^2 \int_0^T \|\omega(t)\|^2, \quad \forall T \geq 0 \tag{12}$$

is satisfied, for $\omega(t) \in L_2 [0,\infty)$, if there exists a dynamic state feedback controller

$$\dot{\Theta} = \begin{cases} -\rho^{-T}\Xi^T L\bar{x}_2 & \text{if } \|\Theta\| < M_\theta \text{ or } (\|\Theta\| = M_\theta \text{ and } \bar{x}_2^T L^T \Xi \Theta \geq 0) \\ -\rho^{-T}\Xi^T L\bar{x}_2 + \rho^{-T}\dfrac{\bar{x}_2^T L^T \Xi \Theta}{\|\Theta\|^2}\Theta & \text{if } \|\Theta\| = M_\theta \text{ and } \bar{x}_2^T L^T \Xi \Theta < 0 \end{cases}$$

$$(13)$$

$$\tau = F_0(x_e) + \Xi\Theta - k_0 E\bar{x}_2 + k(x_e)sgn(L\bar{x}_2) - J^T\lambda_c \quad (14)$$

which is the solution of the adaptive nonlinear H_∞ control problem subject to Equation (6) for the cases 1 and 2.

The stability proof of this dynamic controller follows the line of the proof presented by Chang and Chen [24].

6. FORCE/MOMENT SENSOR DEVELOPMENT

The controllers proposed in Section 5 require the inferred force and moment measures in the robotic end-effector. The most accurate way to obtain such information is by measurements obtained through a multi-axes load cell.

This section presents a detailed description on the modular 3D dynamic force/moment sensor device developed for this application purpose. Its modular architecture is shown in Figure 3.

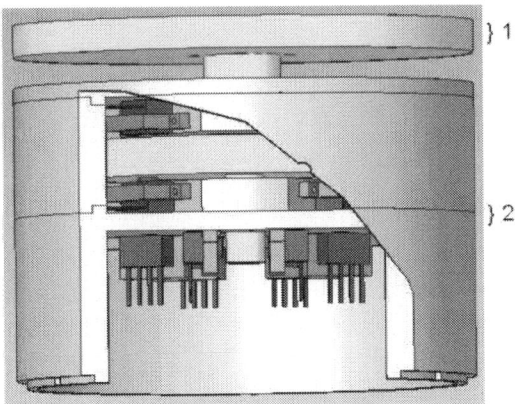

Figure 3. Design of the 3D dynamic forces/moments sensor: (1) Moving part; (2) Fixed part.

6.1. Mechanical Description

The 3D dynamic forces/moments sensor essentially comprised of two parts: a moving part and a fixed part. The moving part is shown in yellow in Figure 3(1). It is composed of a base (outside the sensor body) and a body of force transmission (inside the sensor body), which is responsible for transmitting forces and moments to the unidirectional force sensor units mounted on the fixed part, Figure 3(2). The device has an aluminum-made sensor board where twelve force sensor units are disposed in order to measure force and moment in all directions. The sensor units are arranged in pairs, for example, sensors 1A and 1B as shown in Figure 4, which illustrates the body of forces transmission (4) and identifies the sensor units.

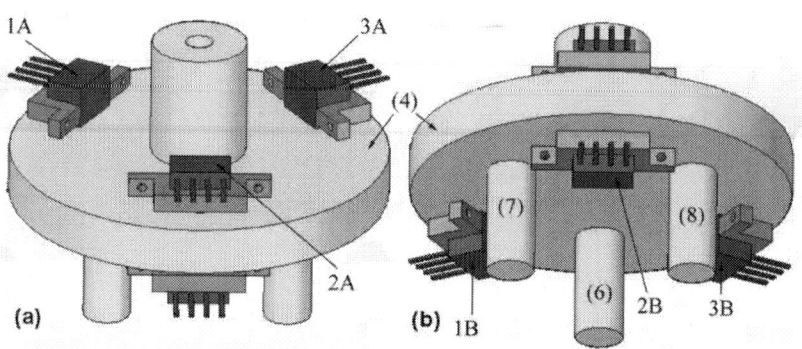

Figure 4. Body of force transmission (4) and arrangement of sensor units: (**a**) Top view; (**b**) Bottom view.

Each pair of sensor units measures the force in one direction. This setting is tensioned proportionally to the movement performed by (4) in that axis. Therefore, forces and moments normal to the plane of the base are calculated by the composition of measured forces in the three axes. Figure 5 demonstrates the sensor units operation schematic, and identifies the force measurements.

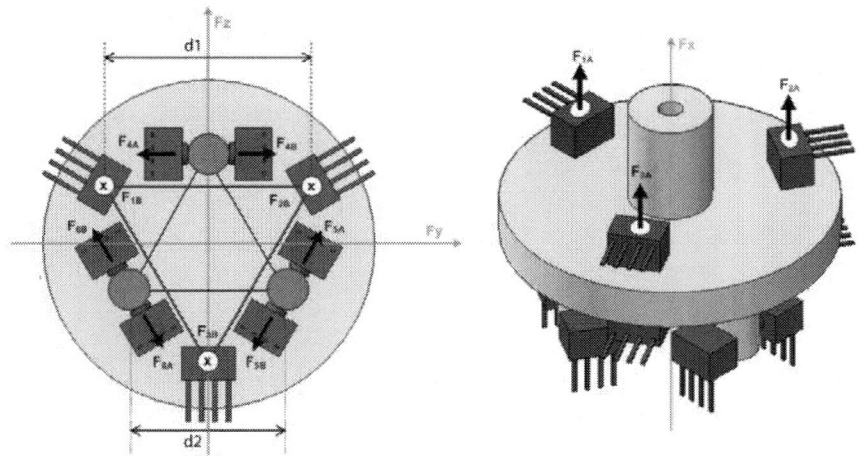

Figure 5. Schematic of sensor operation.

The formulation of the resultant forces/moments composition is obtained as follows.

Three pairs of sensor units, (1A,1B), (2A,2B) and (3A,3B), define, respectively, the force measurements (F_{1A}, F_{1B}), (F_{2A}, F_{2B}) and (F_{3A}, F_{3B}). The resultant forces of these three points can be expressed as

$$F_1 = F_{1A} - F_{1B}$$

$$(15)$$

$$F_2 = F_{2A} - F_{2B}$$

$$(16)$$

$$F_3 = F_{3A} - F_{3B}$$

$$(17)$$

It can be seen in Figure 5 that F_1, F_2, and F_3 form an equilateral triangle, whose side is given by d_1. In [29], Doebelin demonstrated that if three load cells, measuring force in one direction, are arranged triangularly as the vertices of the triangle (the measurement direction normal to the plane formed by the triangle), the following physical quantities can be calculated:

$$F_x = F_1 + F_2 + F_3 \tag{18}$$

$$M_y = d_1 \frac{F_1 - F_2 - (F_3 - F_1)}{2\sqrt{3}} \tag{19}$$

$$M_z = \frac{d_1(F_3 - F_2)}{2} \tag{20}$$

where F_x, M_y, and M_z are the force along x-axis and the moments along axes y and z, respectively.

In the same way, F_4, F_5, and F_6 also form an equilateral triangle, whose side is given by d_2. Thus, the resultant forces of these three points can be expressed as

$$F_4 = F_{4A} - F_{4B} \tag{21}$$

$$F_5 = F_{5A} - F_{5B} \tag{22}$$

$$F_6 = F_{6A} - F_{6B} \tag{23}$$

and therefore,

$$F_y = \frac{F_5 - F_4 - (F_4 - F_6)}{2} \tag{24}$$

$$F_z = \frac{\sqrt{3}(F_5 - F_6)}{2} \tag{25}$$

$$M_x = \frac{-d_2(F_4 + F_5 + F_6)}{2\sqrt{3}} \tag{26}$$

where F_y, F_z, and M_x are the forces along axes y and z and the moment along axis x, respectively. The sensor parts and the assembled device is shown in Figure 6.

The applied sensor unit provides a measurement range of 0 to 1.5 kgf, or a maximum load of 14.7 N. By means of Equations (18)–(20) and (24)–(26), the built sensor presents measurements characteristics as shown in Table 1.

Table 1. Maximum load.

	Axis X	**Axis Y**	**Axis Z**
Force (N)	44.1	29.4	25.5
Moment (N/m)	$d_2$12.7	$d_1$17	$d_1$14.7

Figure 6. 3D dynamic forces/moments sensor.

6.2. Electronic Description

The signal from each pair of unidirectional sensors are conditioned by an electronic circuit based on Instrumentation Amplifiers (INA). The electrical schematic of the signal conditioning circuit is summarized in Figure 7.

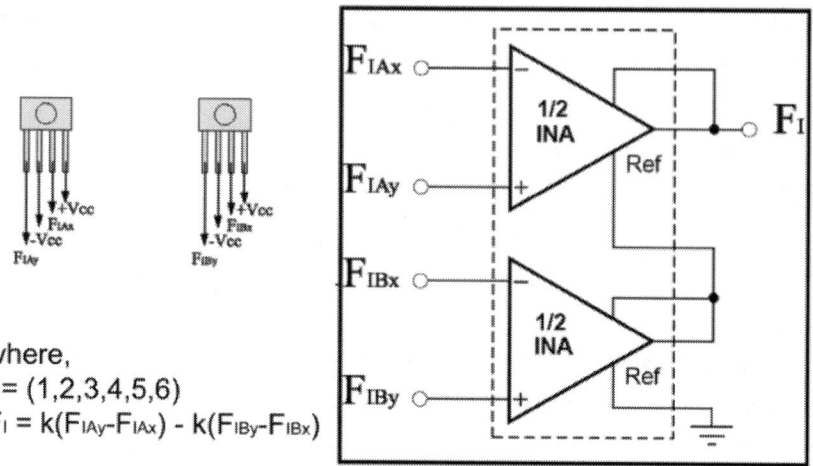

where,
$I = (1,2,3,4,5,6)$
$F_I = k(F_{IAy} - F_{IAx}) - k(F_{IBy} - F_{IBx})$

Figure 7. Electronic diagram.

The sensors are powered by a symmetric supply (+ Vcc and −Vcc) and provide differential output signals (for example, F_{1Ax} and F_{1Ay}).

Six INA comprise the printed circuit board (PCB). Each INA is responsible for conditioning output signals from a pair of force sensors. With this configuration it became possible to measure the differential forces applied on the moving part of the proposed device. The resultant force is then given by the following equation as described in Figure 7.

$$F_I = kF_{IA} - kF_{IB}$$

The signals generated by the INAs are directed to a data acquisition board that interfaces with the computer.

7. EXPERIMENTAL RESULTS

The designed 3D dynamic forces/moments sensor operation is validated in a set of experimental applications, where a three-link planar manipulator is subject to a holonomic constraint defined by a straight rule. The proposed intelligent adaptive robust controllers are applied to this plant for position and force trajectory tracking. The experimental manipulator UARM (UnderActuated Robot Manipulator), whose nominal parameters are given in Table 2, is composed of a DC motor in each joint, a break and an optical encoder with quadrature decoding used to measure joint positions, Figure 8(a). Joint velocities are obtained by numerical differentiation and filtering [30,31]. Modeling matrices for the UARM, M(q), C(q, q˙), and G(q), can be seen in [30]. In Figure 8(b), the designed force sensor is coupled to the UARM end-effector which is constrained in a straight rule. A graphical user interface, whose software was developed using MATLAB* platform, is used to implement control laws and interface with the experimental setup. It was implemented in a modular form, since any new controller can be easily developed and applied.

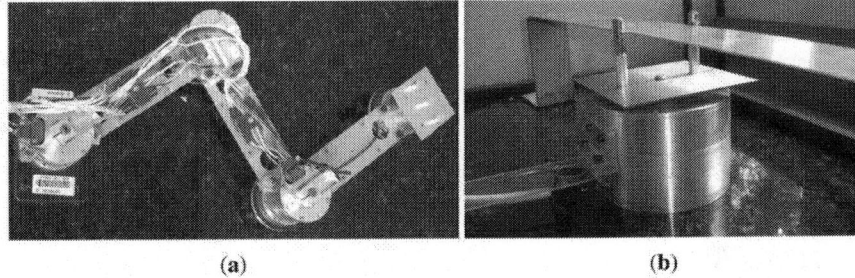

(a)   (b)

Figure 8. (a) UnderActuated Robot Manipulator; (b) Force sensor device coupled to the UARM end-effector.

Table 2. UARM Parameters.

$Body$	m_i	I_i	l_i	lc_i
	(kg)	(kgm^2)	(m)	(m)
Link 1	0.850	0.0075	0.203	0.096
Link 2	0.850	0.0075	0.203	0.096
Link 3	1.700	0.0900	0.240	0.177

7.1. Implementation of Control Law

The constraint surface for the robot end-effector is a segment of a straight line on the X–Y plane. The angle α between the end-effector and the constraint line is defined in this application example as π/2. It means that the orientation must remain in a constant value c given through the line inclination β and α, where $c = \pi + atan(\beta) - \alpha$. Hence, the equation of the m = 2 constraints is given by

$$\phi(q) = \begin{bmatrix} -l_1 s_1 - l_2 s_{12} - l_3 s_{123} + \beta[l_1 c_1 + l_2 c_{12} + l_3 c_{123}] + b \\ q_1 + q_2 + q_3 - c \end{bmatrix} = \begin{bmatrix} 0 \\ 0 \end{bmatrix}$$

where b is the linear coefficient of the constraint line, $s_{ij} = sin(q_i + q_j)$, $c_{ij} = cos(q_i + q_j)$. Hence, $\phi : \Re^3 \rightarrow \Re^2$, and the Jacobian matrix, $J_c(q) = \partial\phi/\partial q$, is given by

$$J_c = \begin{bmatrix} J_{c11} & J_{c12} & J_{c13} \\ J_{c21} & J_{c22} & J_{c23} \end{bmatrix}$$

with

$$J_{c11} = l_1 c_1 + l_2 c_{12} + l_3 c_{123} + \beta[l_1 s_1 + l_2 s_{12} + l_3 s_{123}]$$
$$J_{c12} = l_2 c_{12} + l_3 c_{123} + \beta[l_2 s_{12} + l_3 s_{123}]$$
$$J_{c13} = l_3 c_{123} + \beta[l_3 s_{123}]$$
$$J_{c21} = J_{c22} = J_{c23} = 1$$

Defining $q^1 = [q_1]$ and $q^2 = [q_2 \; q_3]$, the matrix L(q) of the constraint line is

$$L(q) = \begin{bmatrix} 1 \\ -\dfrac{[l_1\cos(q_1)+l_2\cos(q_1+q_2)+\beta(l_1\sin(q_1)+l_2\sin(q_1+q_2))]}{l_2[\cos(q_1+q_2)+\beta\sin(q_1+q_2)]} \\ \dfrac{[l_1\cos(q_1)+l_2\cos(q_1+q_2)+\beta(l_1\sin(q_1)+l_2\sin(q_1+q_2))]}{l_2[\cos(q_1+q_2)+\beta\sin(q_1+q_2)]} - 1 \end{bmatrix}$$

Initial and final coordinates of the movement are $(x_0, y_0) = (0.46, 0.38)$ m and $(x(T), y(T)) = (0.53, 0.13)$ m, respectively. In this case, $\beta = -3.57$, $b = 2.02$, and $c = 15.6°$. The reference trajectory for the joint variables $q_d(t) = q_d^1(t)$ is a fifth-degree polynomial, with trajectory duration time T = 4 s. It is desired that no force acts on the normal direction of the constrain line and no moment acts on the z-direction, that is, $\lambda_d = [(F_x)_d \ \ (M_z)_d]^T = [0 \ \ 0]^T$.

We use as benchmarking for comparison of the proposed controllers, a controller with no intelligent adjust adaptation, that is, only the nominal parameters are used. In other words, we use the controller of Equation (14) without the $\Xi\Theta$ term.

During the experiment, a limited disturbance was introduced at $t_i = 1.0$s in the following form:

$$\tau_d = \begin{bmatrix} 0,01 e^{\frac{-(t-t_d)^2}{2\mu^2}} \operatorname{sen}(3,6\pi t) \\ -0,01 e^{\frac{-(t-t_d)^2}{2\mu^2}} \operatorname{sen}(2,7\pi t) \\ 0,01 e^{\frac{-(t-t_d)^2}{2\mu^2}} \operatorname{sen}(1,8\pi t) \end{bmatrix}$$

If compared with the nominal torque, the disturbance τ_d is approximately 64% of its peak value, see Figure 9.

Figure 9. Torque disturbance.

The selected gains are defined in Table 3.

Table 3. Selected Gains.

p			k_0			k_λ			ρ		
2.50	0	0	0.35	0	0	2.00	0	0	0.75	0	0
0	2.50	0	0	0.35	0	0	2.00	0	0	0.75	0
0	0	2.50	0	0	0.35	0	0	2.00	0	0	0.75

7.1.1. Neural Networks Configuration

For the nonlinear H_∞ controllers via neural network, Cases 1 and 2, proposed in Section 5, let n = 3 be the number of joints of manipulator. Define

$$\hat{F}_{NNi}(x_e, \Theta) = [\ \hat{F}_{NNi_1}(x_e, \Theta_1)\quad \hat{F}_{NNi_2}(x_e, \Theta_2)\quad \hat{F}_{NNi_3}(x_e, \Theta_3)\ ]^T = \Xi\Theta$$

with $p_1 = p_2 = p_3 = 7$ neurons in the hidden layer and the bias vector $b_1 = b_2 = b_3 = [\,-3\ \ -2\ \ -1\ \ 0\ \ 1\ \ 2\ \ 3\,]$. Using the reduced model, the input vector x_e is defined by $xe = [\,q^1\ \ \dot{q}^1\ \ q_d^1\ \ \dot{q}_d^1\ \ \ddot{q}_d^1\,]$. In order to provide an input related to the error values, the weighting matrix for the first layer of the neural networks is defined $W_i^1 = W_i^2 = W_i^3 = [w_{ij}^1] = [w_{ij}^2] = [w_{ij}^3]\ [\,1\ \ 1\ \ -1\ \ -1\ \ -1\,]$. The activation function of the neurons in the hidden layer is chosen as the hyperbolic tangent, $G(.) = tanh(.)$. The uncertain vector Θ is defined as $\Theta = [\,\Theta_1^T\ \ \Theta_2^T\ \ \Theta_3^T\,]^T$ and the matrix $\Xi = [\,\xi_1^T\ \ \xi_2^T\ \ \xi_3^T\,]^T$ as Pazelli et al. and Siqueira et al. showed in [28,30]. Θ values are updated at each control iteration by the adaptive law given in Equation (13). The neural networks outputs are given by

$$
\hat{F}_{NNi_k}(x_e, \Theta_k) = \sum_{i=1}^{p_k} \Theta_{ki} G \left(\sum_{j=1}^{q_k} w_{ij}^k x_{ej} + b_i^k \right) = \xi_k^T \Theta_k
$$

Experimental results for the Case 1 (NN1) with neural network are shown in Figures 10(a), 11(a), and 12(a). For the Case 2 with neural network plus nominal model (NN2), experimental results are shown in Figures 10(b), 11(b), and 12(b).

Remark 2 Although NN1 and NN2 use the same structure and the same input data, their outputs are different. The online adaptation law provides the update of Θ values in distinct ways for each case. The contribution of NN1 output to the final torque signal is lower than the contribution of NN2 output, since NN1 estimates the value of $\Delta F(x_e)$ and NN2 estimates the value of $F(x_e) + \Delta F(x_e)$. Thus, Θ values must achieve that requirement for each case.

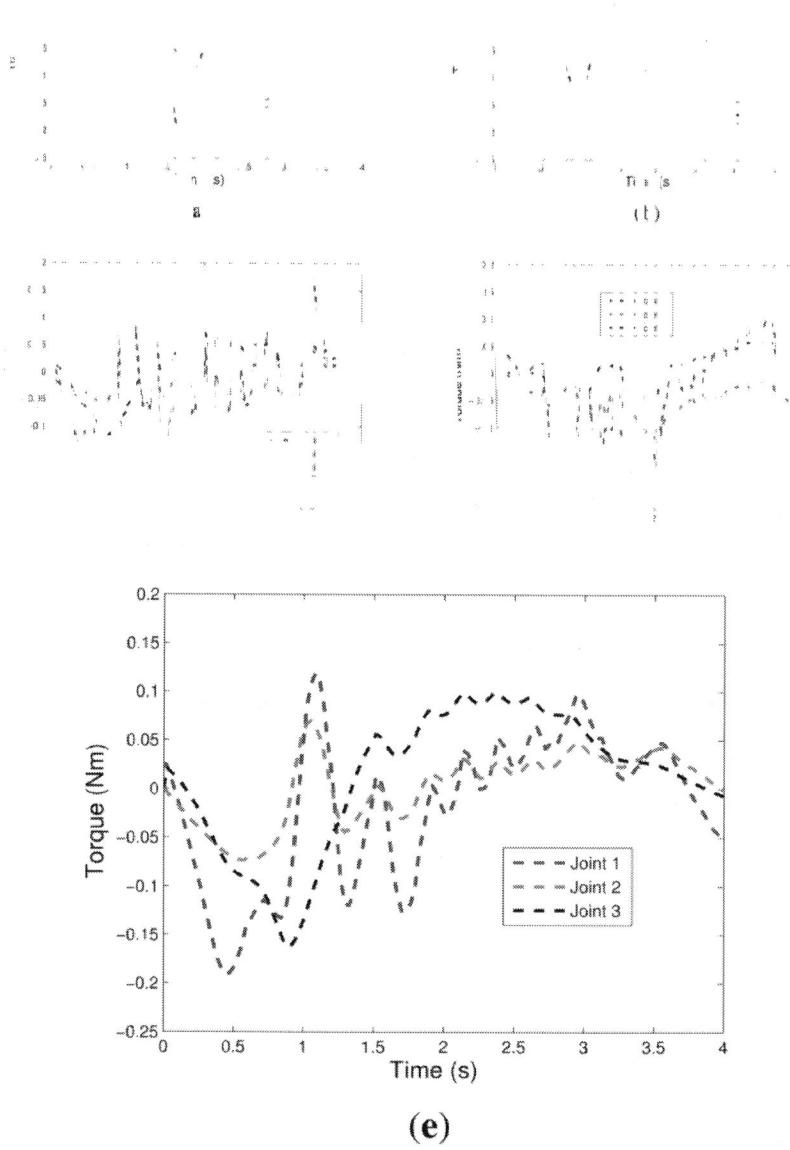

Figure 10. Joint Torque. (**a**) NN1; (**b**) NN2; (**c**) FS1; (**d**) FS2; (**e**) NOM.

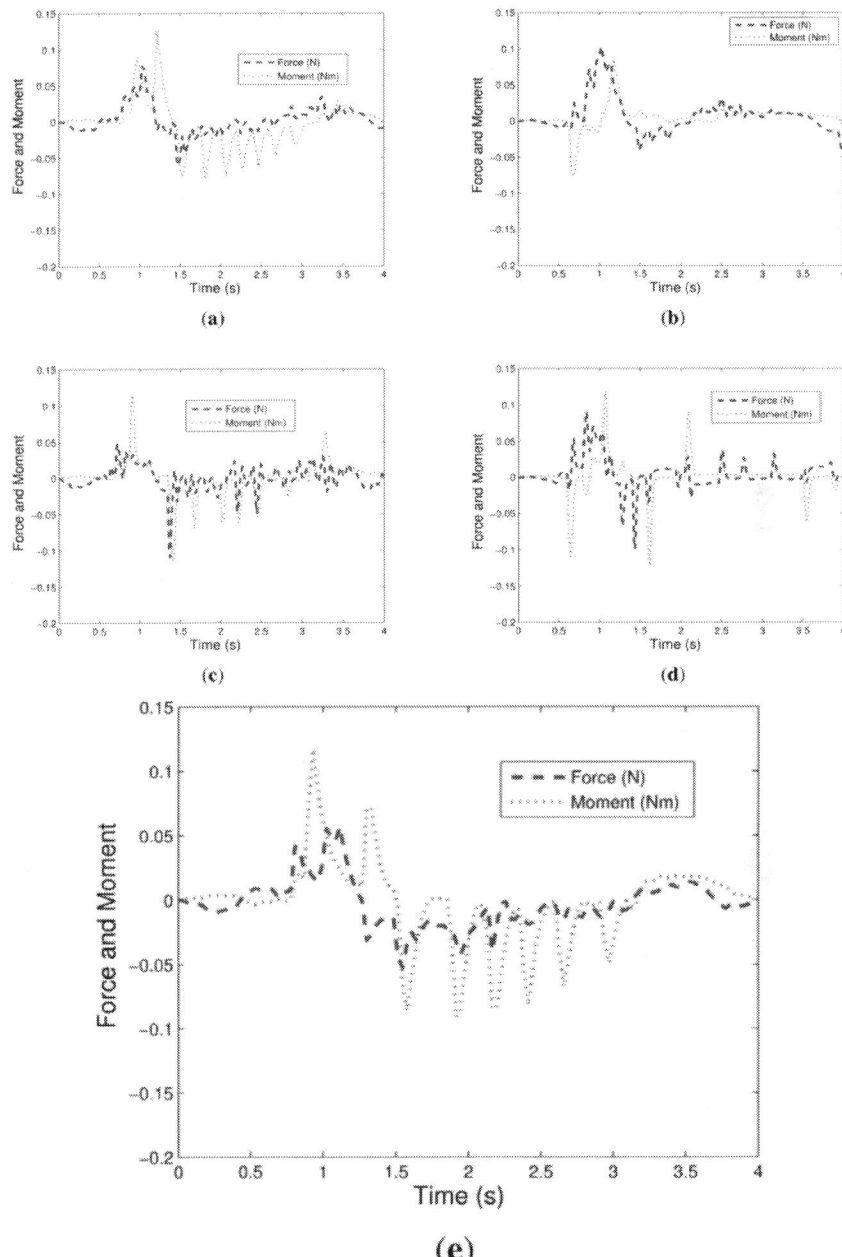

Figure 11. Force and Moment. (**a**) NN1; (**b**) NN2; (**c**) FS1; (**d**) FS2; (**e**) NOM.

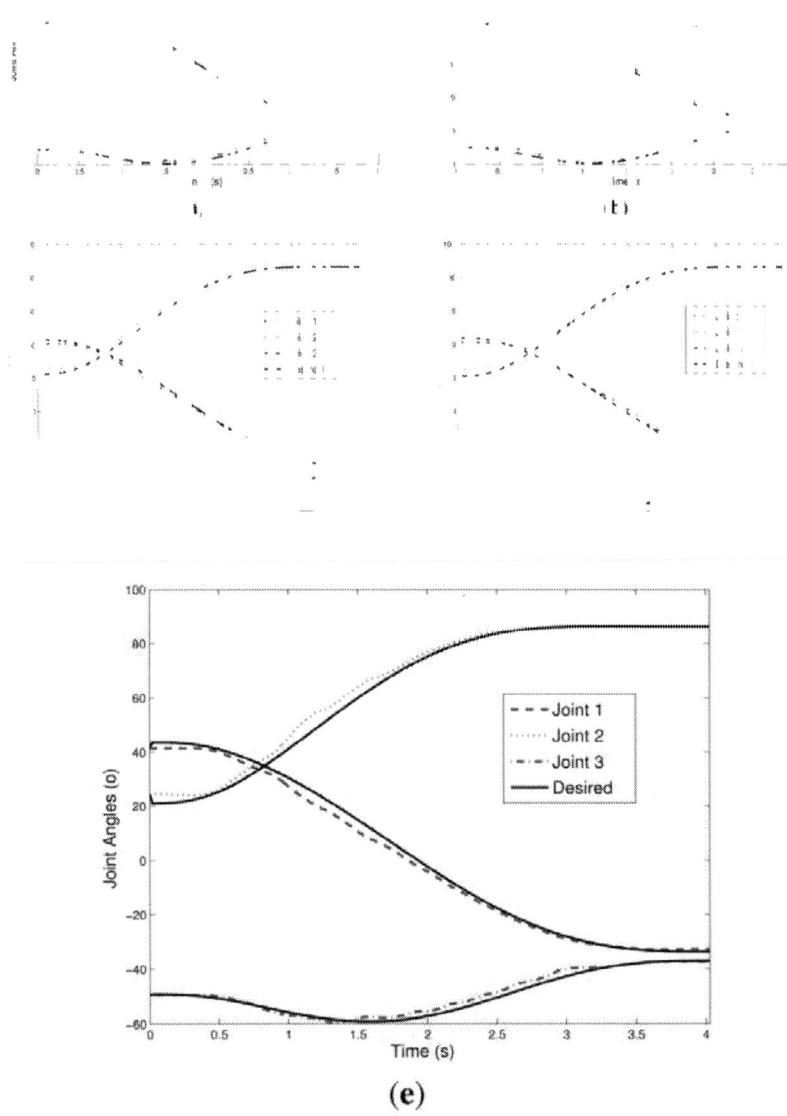

Figure 12. Joint Angles. (**a**) NN1; (**b**) NN2; (**c**) FS1; (**d**) FS2; (**e**) NOM.

7.1.2. Fuzzy Systems Configuration

For the proposed adaptive fuzzy nonlinear H_∞ controllers, Cases 1 and 2 (Section 5), a set of n = 3 fuzzy systems may be defined by

$$\hat{F}_{FSi}(x_e, \Theta) = \left[\begin{array}{ccc} \hat{F}_{FSi_1}(x_e, \Theta_1) & \hat{F}_{FSi_2}(x_e, \Theta_2) & \hat{F}_{FSi_3}(x_e, \Theta_3) \end{array} \right.$$

where $\hat{F}_{FSi_1}(.,.)$, $\hat{F}_{FSi_2}(.,.)$, and $\hat{F}_{FSi_3}(.,.)$ correspond to the estimate of the uncertain part of the dynamic behavior of joints 1, 2, and 3, respectively. The input vector x_e is defined as $x_e = \left[\begin{array}{cc} \tilde{q}_1 & \dot{\tilde{q}}_1 \end{array} \right]$. The fuzzy sets $A(x_e) = \left[\begin{array}{cc} A_1(\tilde{q}_1) & A_2(\dot{\tilde{q}}_1) \end{array} \right]$ are defined according to Figure 13 and the number of linguistic variables in U_1 and U_2 are defined as $r_1 = r_2 = 3$. They are applied to both $\hat{F}_{FSi_1}(.,.)$, $\hat{F}_{FSi_2}(.,.)$ and $\hat{F}_{FSi_3}(.,.)$ for the universe of discourse of position errors, $u_1^1 = u_1^2 = u_1^3 = \tilde{q}_1 \in U_1^1 = U_1^2 = U_1^3 = U_1$, and for the universe of discourse of velocity errors, $u_2^1 = u_2^2 = u_2^3 = \dot{\tilde{q}}_1 \in U_2^1 = U_2^2 = U_2^3 = U_2$.

During the control loop, these graphics are used to determine the grade of membership associated to x_e, which defines the value of μ in the output of the Takagi–Sugeno fuzzy model; for more details see [28],

$$\hat{F}_{FSi_k}(x_e, \Theta_k) = \frac{\sum_{i=1}^{p_k} \mu_i^k y_i^k}{\sum_{i=1}^{p_k} \mu_i^k} = \frac{\sum_{i=1}^{p_k} \mu_i^k (\theta_{i0}^k + \theta_{i1}^k u_1 + \theta_{i2}^k u_2)}{\sum_{i=1}^{p_k} \mu_i^k} = \xi_k^T \Theta_k$$

A fuzzy rule base is defined with $p_k = r_1 r_2 = 9$ rules as

$$R_i : \textbf{IF}(u_1 \text{ is } A_{r_1 i 1}) \text{ and } (u_2 \text{ is } A_{r_2 i 2}) \textbf{ THEN } y_i$$

The vector of adjustable components Θ is defined as $\Theta = [\ \Theta_1^T\ \ \Theta_2^T\ \ \Theta_3^T\ ]^T$, with

$$\Theta_1^T = [\ \theta_{10}^1\ \ \theta_{11}^1\ \ \theta_{12}^1\ \cdots\ \theta_{90}^1\ \ \theta_{91}^1\ \ \theta_{92}^1\ ]$$
$$\Theta_2^T = [\ \theta_{10}^2\ \ \theta_{11}^2\ \ \theta_{12}^2\ \cdots\ \theta_{90}^2\ \ \theta_{91}^2\ \ \theta_{92}^2\ ]$$
$$\Theta_3^T = [\ \theta_{10}^3\ \ \theta_{11}^3\ \ \theta_{12}^3\ \cdots\ \theta_{90}^3\ \ \theta_{91}^3\ \ \theta_{92}^3\ ]$$

and are updated at each control iteration by the adaptive law in Equation (13).

The matrix $\Xi = [\ \xi_1^T\ \ \xi_2^T\ \ \xi_3^T\ ]^T$ is computed with

$$\xi_1^T = [\ \xi_{10}^1\ \ \xi_{11}^1\ \ \xi_{12}^1\ \cdots\ \xi_{90}^1\ \ \xi_{91}^1\ \ \xi_{92}^1\ ]$$
$$\xi_2^T = [\ \xi_{10}^2\ \ \xi_{11}^2\ \ \xi_{12}^2\ \cdots\ \xi_{90}^2\ \ \xi_{91}^2\ \ \xi_{92}^2\ ]$$
$$\xi_3^T = [\ \xi_{10}^3\ \ \xi_{11}^3\ \ \xi_{12}^3\ \cdots\ \xi_{90}^3\ \ \xi_{91}^3\ \ \xi_{92}^3\ ]$$

Figures 10(c), 11(c), and 12(c) show the experimental results for Case 3. Figures 10(d), 11(d), and 12(d) show the experimental results of fuzzy system plus nominal model for Case 4.

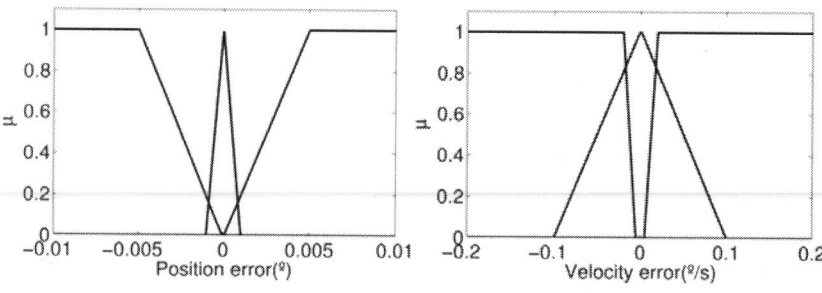

Figure 13. Fuzzy sets $A_1(\tilde{q}_1)$ and $A_2(\tilde{q}_1)$.

7.2. Results Discussion

Figure 10 shows the applied torques in the robotic manipulator joints. Note that for the range of the disturbance inserted, the controllers act strongly on the system, reversing the directions of the torques of joints 1 and 2 and increasing the torque of joint 3. In this analysis, it appears that the controllers using neural networks, NN1 and NN2, oscillate less, therefore, they have a slightly slower response time.

The charts of Figure 11 are important because they show the behavior of the end effector during the controller actions, and represent the measurements of force and moment in the robotic end-effector. As expected, it was observed that during the period in which the disturbance appeared in the system, there was a higher intensity of forces and moments which decreased gradually, tending to zero, until the final time (4 s). The proposed configuration for the device sensor proves its effectiveness in providing these data with the required accuracy and without delay to the implementation of the control laws under study.

Figure 12 shows the tracking angles of three joints of the robotic manipulator. Notice that the largest deviation is in the range where the disturbance appeared. Nonetheless, the experiment showed good tracking of the trajectories desired. Again, we can see that for the controller without the intelligent system, NOM, the control actuation is slower, providing greater tracking error of reference signals.

Three performance indexes are used to compare the nonlinear H_∞ controllers: the L_2 norm of the state vector

$$\mathcal{L}_2[\tilde{x}] = \left(\frac{1}{(t_r - t_0)} \int_{t_0}^{t_r} \|\tilde{x}(t)\|_2^2 \, dt \right)^{1/2}$$

where $\| \cdot \|_2$ is the Euclidean norm, the sum of the applied torques

$$E[\tau] = \sum_{i=1}^{3} \left(\int_{t_0}^{t_r} |\tau_i(t)| \, dt \right)$$

and the sum of areas of contact forces

$$E[\lambda] = \sum_{i=1}^{3} \left(\int_{t_0}^{t_r} |\lambda_i(t)| \, dt \right)$$

where $\lambda_i(t)$ the ith component of the contact forces. As the desired values of contact forces are zero, the lower the value of E $[\lambda]$, the better the controller will be with respect to contact force control.

The results that are shown in Table 4 present the mean value of five experiments.

Table 4. Performance Indexes.

	$\mathcal{L}_2[x]$ (m)	$E[\tau]$ (Nms)	$E[\lambda]$ (Ns)
Nominal	0.1031	0.6508	0.1506
NN1	0.0780	0.6564	0.1435
NN2	0.1026	0.4318	0.1156
FS1	0.0759	0.5865	0.1136
FS2	0.0899	0.5182	0.1061

From Table 4 we can conclude that the controller based only on the nominal model presents a higher state error and greater forces. Although the fuzzy controllers exhibit the best performance, this performance difference with respect to controllers based on neural networks deserves more in-depth study, which will be carried out in future works.

8. CONCLUSIONS

In this work, the control problem of trajectory tracking with H_∞ guaranteed performance was considered for manipulators with force and position constraint. Five controllers were evaluated:

- The first controller, called nominal, considers that the term $F(x_e)$ is completely known, i.e., it does not take into account parametric uncertainties of the model.

- The second and third controllers have neural network-based estimators, where the second considers the known nominal model of the manipulator and estimates only parameter uncertainties, and the third estimates the full model.

- The fourth and fifth controllers have fuzzy logic-based estimators which estimate only parametric uncertainties and the full model, respectively.

According to experimental results presented in Section 7, controllers NN1 and FS1 have better tracking trajectory compared, respectively, with the NN2 and FS2. Moreover, it also appears they are more stable because of the lower oscillations after the introduction of disturbance, and this feature occurs for the trajectory tracking in Cartesian space in order to follow the trajectory of joint angles. Comparing the results of intelligent systems it is noted that the controllers via neural networks (NN2 and NN1) have a smoother control action and, consequently, have smaller oscillations than the fuzzy systems (FS1 and FS2), which can be verified by comparing the graphics of torque (Figure 10). On the other hand, the controllers NN1 and NN2 are slower than the FS1 and FS2, hence they tend to have a greater tracking error. In addition, when the system is subject to disturbances, all controllers respond well.

Analyzing the proposed performance indices, it was shown that the state error in the fuzzy system-based controllers (FS1 and FS2) tend to be smaller than those based on neural networks (NN2 and NN1). A possible explanation for this is due to the fact that fuzzy systems acted faster than neural networks. It was observed that choosing values of k_λ more directly influence the adjustment of the forces, while the values of ρ influence the tracking errors in the state variables.

The development of the proposed sensor was of crucial importance for the realization of the experimental analysis, a differential of this paper. The measurement results shown proved the effectiveness of the proposed structure. Still, its modular feature enables the application of sensor units of various kinds, such as piezoelectric sensors.

As future work, a comparative study will be performed between the developed sensor and commercially available ones. Furthermore, the use of genetic algorithms to adjust controller gains is a current and ongoing project.

ACKNOWLEDGMENTS

This work was supported by Coordenação de Aperfeiçoamento de Pessoal de Nível Superior (CAPES) and Fundação de Apoio à Pesquisa do Estado de São Paulo (FAPESP), under grant 2012/05552-9.

REFERENCES

1. Kilicaslan, S.; Özgören, M.K.; Ider, K. Hybrid force and motion control of robots with flexible links. Mech. Mach. Theory **2010**, 45, 91–105.

2. Lewis, F.L.; Abdallah, C.T.; Dawson, D.M. Control of Robot Manipulators; MacMillan Publishing Company: New York, NY, USA, 1993.

3. Bechlioulis, C.P.; Doulgeri, Z.; Rovithakis, G.A. Neuro-adaptive force/position control with prescribed performance and guaranteed contact maintenance. IEEE Trans. Neural Netw. **2010**, 21, 1857–1868.

4. Cheah, C.C.; Hou, S.P.; Zhao, Y.; Slotine, J.J.E. Adaptive vision and force tracking control for robots with constraint uncertainty. IEEE/ASME Trans. Mechatron. **2010**, 15, 389–399.

5. Karayiannidis, Y.; Doulgeri, Z. Adaptive control of robot contact tasks with on-line learning of planar surfaces. Automatica **2009**, 45, 2374–2382.

6. Karayiannidis, Y.; Doulgeri, Z. Robot contact tasks in the presence of control target distortions. Robot. Auton. Syst. **2010**, 58, 596–606.

7. Kumar, N.; Panwar, V.; Sukavanam, N.; Borm, S.P.S.J.H. Neural network based hybrid force/position control for robot manipulators. Int. J. Precis. Eng. Manuf. **2011**, 12, 419–426.

8. Salisbury, J.K. Active Stiffness Control of a Manipulator in Cartesian Coordinates. Proceedings of the 19th IEEE Conference on Decision and Control including the Symposium on Adaptive Processes, Albuquerque, NM, USA, 10–12 December 1980; Volume 19, pp. 95–100.

9. Raibert, M.; Craig, J. Hybrid position/force control of manipulators. ASME J. Dyn. Syst. Meas. Control **1981**, 102, 126–133.

10. Paul, R.; Shimano, B. Compliance and Control. Proceedings of the Joint Automatic Control Conference, Purdue, IN, USA, 27–30 July 1976; pp. 694–699.

11. McClamroch, N.H.; Wang, D. Feedback stabilization and tracking of constrained robots. IEEE Trans. Autom. Control **1988**, 33, 419–426.

12. Ortega, R.; Spong, M.W. Adaptive motion control of rigid robots: A tutorial. Automatica **1989**, 25, 877–888.

13. Slotine, J.J.E.; Li, W. Composite adaptive control of robot manipulators. Automatica **1989**, 25, 509–519.

14. Slotine, J.J.E.; Li, W. Applied Nonlinear Control; Prentice-Hall: Englewoods Cliffs, NJ, USA, 1991.

15. Spong, M.W.; Vidyasagar, M. Robot Dynamics and Control; John Wiley and Sons: New York, NY, USA, 1989.

16. Carelli, R.; Kelly, R. An adaptive impedance/force controller for robot manipulators. IEEE Trans. Autom. Control **1991**, 36, 967–971.

17. Jean, J.H.; Fu, L.C. Adaptive hybrid control strategies for constrained robots. IEEE Trans. Autom. Control **1993**, 38, 598–603.

18. Panteley, E.V.; Stotsky, A.A. Adaptive trajectory/force control scheme for constrained robot manipulators. Int. J. Adapt. Control Signal Process. **1993**, 7, 489–496.

19. Su, C.Y.; Stepanenko, Y.; Leung, T.P. Combined adaptive and variable structure control for constrained robots. Automatica **1995**, 31, 483–488.

20. Su, C.Y.; Leung, T.P.; Zhou, Q.J. Force/motion control of constrained robots using sliding mode. IEEE Trans. Autom. Control **1992**, 37, 668–672.

21. Chang, Y.C.; Chen, B.S. Adaptive tracking control design of constrained robot systems. Int. J. Adapt. Control Signal Process. **1998**, 12, 495–526.

22. Chang, Y.C.; Chen, B.S. A nonlinear adaptive H_∞ tracking control design in robotic systems via neural networks. IEEE Trans. Control Syst. Technol. **1997**, 5, 13–28.

23. Chang, Y.C. Neural network-based H_∞ tracking control for robotic systems. IEEE Proc. Control Theory Appl. **2000**, 147, 303–311.

24. Chang, Y.C.; Chen, B.S. Robust tracking designs for both holonomic and nonholonomic constrained mechanical systems: Adaptive fuzzy approach. IEEE Trans. Fuzzy Syst. **2000**, 8, 46–66.

25. Meyer, R.A.; Lowe, A.E. Multiple Axis Load Sensitive Transducer. U.S. Patent 4640138, 1987.

26. Meyer, R.A.; Olson, D.J. Load Transducer. U.S. Patent 4821582, 1989.

27. Sommerfeld, J.L.; Meyer, R.A.; Larson, B.A.; Olson, D.J. Multi-Axis Load Cell. U.S. Patent 5969268, 1999.

28. Pazelli, T.F.P.A.T.; Terra, M.H.; Siqueira, A.A.G. Experimental investigation on adaptive robust controller designs applied to a free-floating space manipulator. Control Eng. Pract. **2011**, 9, 395–408.

29. Doebelin, E.O. Measurement Systems Application and Design; McGraw-Hill Science: New York, NY, USA, 1989; pp. 410–413.

30. Siqueira, A.A.G.; Terra, M.H.; Bergerman, M. Robust Control of Robots: Fault Tolerant Approaches; Springer: London, UK, 2011.

31. Siqueira, A.A.G.; Terra, M.H. Nonlinear and Markovian H_∞ controls of underactuated manipulators. IEEE Trans. Control Syst. Technol. **2004**, 12, 811–826.

CHAPTER 7

A Method for Improving the Pose Accuracy of a Robot Manipulator Based on Multi-Sensor Combined Measurement and Data Fusion

Bailing Liu, Fumin Zhang [*] and Xinghua Qu

State Key Laboratory of Precision Measuring Technology and Instruments, Tianjin University, Tianjin 300072, China

ABSTRACT

An improvement method for the pose accuracy of a robot manipulator by using a multiple-sensor combination measuring system (MCMS) is presented. It is composed of a visual sensor, an angle sensor and a series robot. The visual sensor is utilized to measure the position of the manipulator in real time, and the angle sensor is rigidly attached to the manipulator to obtain its orientation. Due to the higher accuracy of the multi-sensor, two efficient data fusion approaches, the Kalman filter (KF) and multi-sensor optimal information fusion algorithm (MOIFA), are used to fuse the position and orientation of the manipulator. The simulation and experimental results show that the pose accuracy of the robot manipulator is improved dramatically by 38%~78% with the multi-sensor data fusion.

Comparing with reported pose accuracy improvement methods, the primary advantage of this method is that it does not require the complex solution of the kinematics parameter equations, increase of the motion constraints and the complicated procedures of the traditional vision-based methods. It makes the robot processing more autonomous and accurate. To improve the reliability and accuracy of the pose measurements of MCMS, the visual sensor repeatability is experimentally studied. An optimal range of $1 \times 0.8 \times 1 \sim 2 \times 0.8 \times 1$ m in the field of view (FOV) is indicated by the experimental results.

KEYWORDS

pose accuracy; multi-sensor data fusion; visual sensor; series robot

1. INTRODUCTION

Since the first demonstration by Devol et al. in 1956, robots have been widely exploited in many fields, such as spraying, painting, spot welding, sealing, parts picking and other operations [1]. A fine accuracy in terms of the robot manipulator pose is mostly required in recent applications of industrial robots. The pose accuracy is defined precisely by ISO 9283 (1998) [2] as the deviation that occurs between the required and attained poses and the variance of the attained poses in a number of repetitions [1].

Considerable research for many years has been done about the improvement of the pose accuracy of robots. The kinematics errors refer to the differences between the kinematics parameters of a robot and its nominal values because of the manufacturing and assembly tolerance. Cost restriction aside, the kinematics calibration is an effective method to improve the absolute accuracy of robots [3]. A number of research of the kinematics calibration has been presented, e.g., Denavit and Hartenberg (1957) proposed a D-Hmodel that provided the basis for the kinematics calibration. Further, Hayati (1983) presented a revised D-H model for proposing a linear model, which related the parameter errors to the end-effector positioning error of the serial robot directly. However, these methods have limitations. For example, due to both the geometric errors and non-geometric errors, the kinematics model used in the robot controller cannot accurately describe the kinematics transformation of the actual robot. This will result in a large positioning inaccuracy [2]. Moreover, the calibration is usually performed off line, but the kinematics parameter errors often change with the load or environment variation. Therefore, the online and independent measurement is indispensable to improve the pose accuracy.

To avoid these disadvantages, some researchers begun trying to improve the pose accuracy from the perspective of the robot external measurement. They obtain the robot pose through external sensors to directly monitor the tool, workpiece and manipulator instead of considering the kinematics parameters and the influence caused by the environment. Frank S. Cheng (2007) proposed a method of robot cell calibrations to recover the accuracy of the originally defined robot tool-center-point (TCP) positions by employing a precise external sensor measuring system [4]. Hans de Ruiter (2008) presented a 3D-model-based computer-vision method for tracking the full six DOF pose of a rigid body in real time via a combination of the textured model projection and the optical flow [5]. Kaijen Hsiao (2011) applied a robot hand with tactile sensors, to localize the object on a table and ultimately achieved a target placement [6]. Qu Weiwei (2011) presented a closed-loop tracking system based on a laser sensor to reduce the relative pose error of the robot to less than 0.2 mm and $\pm 1''$ in the robot-aided aircraft assembly drilling process [7]. Guanglong Du (2014) presented an online robot self-calibration method that utilized a position sensor to obtain the position of the manipulator and an inertial measurement unit (IMU) to obtain the orientation of the manipulator in real time [3]. However, these methods also have limitations. The traditional vision-based methods utilized to calibrate a robot require the special complex steps, such as the camera calibration, corner detection and laser alignment. The laser-based methods require a large and open-sided space, and the laser beam is easily sheltered during the motion of the robot manipulator. These procedures are inconvenient, time consuming and infeasible for some applications. Therefore, in this paper, we develop a novel and flexible pose measurement system of a robot based on a visual sensor and an angle sensor. This is a quick and efficient method to improve the pose accuracy through fusing the redundant data of the multi-sensor. Our method does not require the complex solution of the kinematics parameter equations and the complicated procedures of the traditional vision-based methods, such as the camera calibration, corner detection and laser alignment. Moreover, this method is not influenced by the load and environmental variations with the online measurement. Those characteristics make the robot processing more autonomous, efficient and accurate.

In this work, we construct the flexible pose measurement system by a dynamic three-dimensional photogrammetry system, a high precision digital inclinometer and a six DOF series robot, which is rarely seen in similar research, so far. It is generally known that the absolute pose accuracy of the robot is worse than its repeatability accuracy. Fortunately, the combination of the high-precision sensors can improve the absolute pose accuracy. In this paper, as a non-contact sensor, the photogrammetry is an accurate technique that is widely used in industrial settings, yielding a measurement precision up to 1:200,000 [8]. The angle sensor, the inclinometer, has a high sensitivity to minor variations of the angle as small as 0.01°. Together with the high-precision encoding of the robot itself, the combination of all three sensors could compensate for each other to achieve a higher pose accuracy ultimately. The multi-sensors will generate a mass of data. For fusing the redundant data, this paper presents two kinds of data fusion methods, the Kalman filter fusion method (KF) and multi-sensor optimal information fusion algorithm (MOIFA), which are modeled and applied in this research.

Besides the factor of the robot itself, the measuring error of the visual sensor also can cause the inaccuracy of results. Therefore, we take the repeatability error of the visual sensor as a research object, and we find an optimal field of view (FOV) of the photogrammetry system in which its repeatability accuracy is the best.

2. METHOD FOR THE IMPROVEMENT OF THE POSE ACCURACY OF THE ROBOT

2.1. System Constitution

The conventional robot system cannot decrease the pose error at the control level due to the open-loop control architecture and simplified control laws. Instead of open-loop robot systems, the multiple-sensor combination measuring system (MCMS) presented in this paper sets up a closed-loop measurement system to improve the pose accuracy of the robot. As shown in Figure 1, the described system is made up of a series robot, an industrial photogrammetry system, a digital inclinometer and the PC software. As we know, the excellent measurement performance is achieved through the use of highly sophisticated components. In this paper we adopt a 6 DOF robot, whose repeatability accuracy is 10^{-3} mm. This robot is offered by KUKA Co., Ltd. (Shanghai, China) and its model is KP 5 arc. We also choose a high precision industrial three-dimensional photogrammetry system to dynamically track and measure the robot pose in real time. The photogrammetry system is composed of 4 motion-sensitive CCD cameras set on top of the robot. We also adopt a high accuracy digital inclinometer LE-30, whose accuracy is 0.01°. The inclinometer can rapidly measure both the angle of the pitch and yaw at a 20-ms frequency, and it can be set up on the robot manipulator to measure the attitude angle in real time. However, only possessing highly sophisticated instruments is not enough; the pose accuracy will be improved through fusing the redundant data of the sensors. These data should be converted by the coordinate transformation matrix previously.

2.2. Method of Data Fusion

As is well known, the primary aim of the sensor fusion is to improve the accuracy by using the redundant information gathered from multiple sources [9]. Since the construction of the MCMS, the sensor fusion is suitable to apply in this system. Currently, a number of different types of data fusion methods are being used in industry, such as the Wiener filter, constrained least squares filter, $\alpha\text{-}\beta$ filter, $\alpha\text{-}\beta\text{-}\gamma$ filter, Kalman filter [10], linear minimum variance fusion algorithm, etc. The Kalman filter (KF), first proposed in 1960 by Kalman [11], has been successfully used in the Apollo moon flight and C-5A aircraft navigation systems. Walker and Harries [12] improved the system robustness and adaptability in the mobile robot area through KF and multi-sensor fusion. The multi-sensor optimal information fusion algorithm in the linear minimum variance sense (MOIFA) is a geometric fusion method that was developed by Nakamura [13] and has been enhanced by Elliot et al. [14]. A demonstration of its use can be seen in [15], where the method was applied to an optical encoder and a camera sensor [16]. The two methods allow improving the fusion accuracy significantly, so we choose both of them as the methods of data fusion in this paper. We will summarize them respectively in Sections 2.2.1 and 2.2.2.

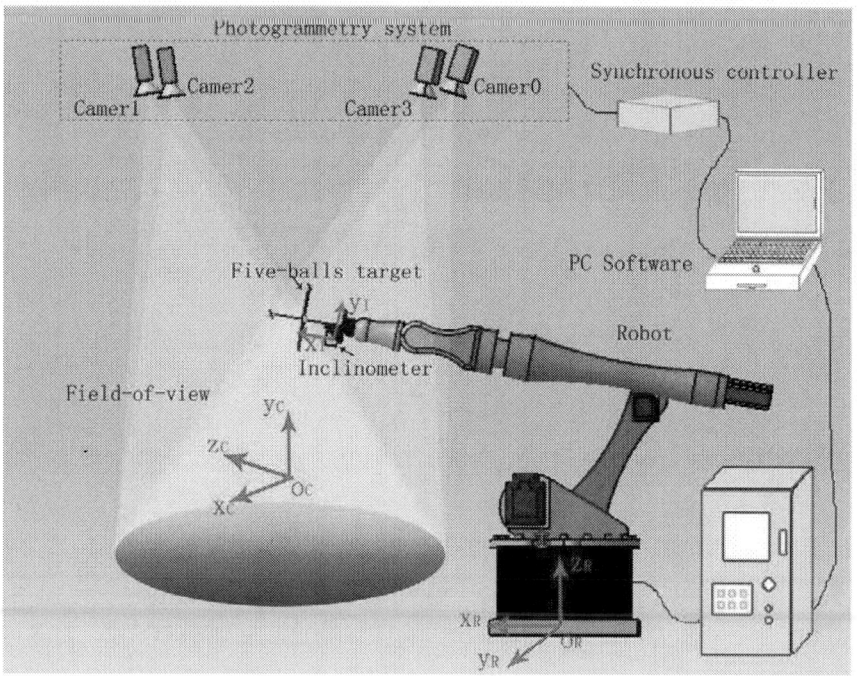

Figure 1. The schematic diagram of the multiple-sensor combination measuring system (MCMS).

2.2.1. Kalman Filter

The Kalman filter solves the optimal linear filtering problems on the basis of a minimum mean square error method. The present value of the signal can be calculated according to the prior predicted value and the latest observation data. The Kalman filter predicts the value through a group of state equations and recursive methods. This recursive solution usually is expressed in the form of the predicted value. The following Equations (1)–(5) are the recursion formulas of the Kalman filter:

$$X(k) = X(k/k - 1) + H(k)[Y(k) - C(k)X(k/k - 1)] \tag{1}$$

$$X(k/k - 1) = A(k)X(k - 1) \tag{2}$$

$$H(k) = p(k/k - 1)C(k)^T[C(k)p(k/k - 1)C(k)^T + R(k)]^{-1} \tag{3}$$

$$p(k/k - 1) = A(k)p(k - 1)A(k)^T + B(k)Q(k)B(k)^T \tag{4}$$

$$p(k) = [I - H(k)C(k)]p(k/k - 1) \tag{5}$$

where X (k) is a multi-dimensional state vector i.e., the predicted value at time k of a single sensor. C (k) is an observation vector. A (k) and B (k) are the transfer matrices determined by the system. Y (k) is the observation value of a single sensor. Q (k) is the system noise matrix, and R (k) is the measurement noise matrix. The statistics features E $[X(0)]$ and var $[X(0)]$ of the initial state $X(0)$ are known as $X(0) = E[X(0)] = \mu_0$ and $p(0) = E[(X(0) - E(X(0)))^2] = \text{var}[X(0)]$.

Substituting p(0) into Equation (4), we obtain $p(1/0)$. Substituting $p(1/0)$ into Equation (3), we obtain $H(1)$. Substituting H (1) into Equation (1), we obtain X (1) in the condition of the minimum mean square error. At the same time, substituting p (1/0) into Equation (5), we obtain p (1). Then, we obtain p (2/1) by p (1), H (2) by p (2/1) and X (2) by H (2), the as same as above, and so on. Therefore, the predicted value at time k can be calculated.

In this paper, we only take into account the position accuracy of the robot. Two sensors, the photogrammetry system and the series robot, are utilized. $Y^1(k)$ is the observation value at time k of the photogrammetry system. $Y^2(k)$ is the observation value at time k of the robot, and $X^1(k)$ is the predicted value at time k of the photogrammetry system. $X^2(k)$ is the predicted value at time k of the robot. $X^f(k)$ is fused by $X^1(k)$ and $X^2(k)$ with the weighting matrix W, which is determined by the typical accuracy values of the measuring instruments. Following is the principle of data fusion:

X^i is the three-dimensional coordinate of the predicted value with the i-th sensor. It usually can be expressed by means of a function of θ as $X^i = f(\theta^i)$, and θ is a multi-dimensional vector. A predicted value with additive noise can be represented as $X^i = f(\theta^i + \delta\theta^i)$, where $\delta\theta^i$ represents the additive noise. Equation (6) is deduced by Taylor expanding X^i and neglecting the quadratic term.

$$X^i = f(\theta^i) + J(\theta^i)\delta\theta^i$$

(6)

where J(θ^i) is the Jacobian matrix of the i-th sensor, $J(\theta^i) = \frac{\partial X^i}{\partial \theta^i}$.

Assuming a Gaussian distribution for the noise gives $E[\delta\theta^i] \triangleq \overline{\delta\theta^i} = 0$. Combining Equation (6), the mean and covariance of X^i are:

$$E[X^i] \triangleq \overline{X^i} = f(\theta^i)$$

(7)

$$V[X^i] \triangleq E[(X^i - \overline{X^i})(X^i - \overline{X^i})^T] = E[J^i\delta\theta^i\delta\theta^{iT}J^{iT}] = J^iQ^iJ^{iT}$$

(8)

where Q^i is the covariance matrix of $\delta\theta^i$.

The weight matrix is defined as $W = (W^1 W^2 ... W^n)$, and n is the number of the sensors. The fusion value at time k combines multiple measurements by the weighted average:

$$X(k) = \sum_{i=1}^{n} W^i X^i(k)$$

$$(9)$$

where $W \in 3 \times 3n$. W^i is the weighting matrix of the i-th sensor. It is assumed that $\overline{X^i} = \overline{X}$, i.e., all measurement instruments are properly calibrated. Using Equations (9) and (7), the fused mean value becomes:

$$E[X(k)] = \sum_{i=1}^{n} W^i E[X^i(k)] = \sum_{i=1}^{n} W^i \overline{X^i}(k) = \left(\sum_{i=1}^{n} W^i \right) \overline{X}(k)$$

$$(10)$$

since,

$$E[X(k)] = \overline{X}(k)$$

$$(11)$$

then, $\sum_{i=1}^{n} W^i = I$, where I is the identity matrix.

Using Equations (6)–(11), the covariance matrix becomes:

$$V[X(k)] = E[(X(k) - \overline{X}(k))(X(k) - \overline{X}(k))^T]$$
$$= E\left[\left(\sum_{i=1}^{n} W^i X^i(k) - \sum_{i=1}^{n} W^i \overline{X^i}(k) \right) \left(\sum_{i=1}^{n} W^i X^i(k) - \sum_{i=1}^{n} W^i \overline{X^i}(k) \right)^T \right]$$
$$= \sum_{i=1}^{n} W^i J^i Q^i J^{iT} W^{iT}$$

$$(12)$$

W^i can be solved by means of Lagrange's method. The solving process is detailed in [13]. The weighting matrix is given to be:

$$W^i = \left\{ \sum_{i=1}^{n} \left(J^i Q^i J^{iT} \right) \right\}^{-1} \left(J^i Q^i J^{iT} \right)^{-1} \qquad (13)$$

In this paper, W^1 and W^2 represent the weighting matrices of the photogrammetry system and the series robot separately. According to Equation (9), the fused result at time k is:

$$X^f(k) = W^1 X^1(k) + W^2 X^2(k) \qquad (14)$$

2.2.2. Multi-Sensor Optimal Information Fusion Algorithm

The optimum fused value of the spatial coordinates for the robot position can be calculated as shown in Equation (15). "The optimum value" means the minimum variance unbiased estimate of the fused result X.

$$\hat{X} = \sum_{i=1}^{n} W^i X^i \qquad (15)$$

where i is the number of measuring instruments. The weighting matrix is shown to be:

$$W^i = \left\{ \sum_{i=1}^{n} \left(J^i Q^i J^{iT} \right) \right\}^{-1} \left(J^i Q^i J^{iT} \right)^{-1} \qquad (16)$$

where Q^i, J^i are the covariance matrix and the Jacobian matrix of the i-th measuring instrument. In this paper, the covariance matrices of the photogrammetry system and robot are shown below:

$$Q^c = \begin{bmatrix} \delta_{xc}^2 & & \\ & \delta_{yc}^2 & \\ & & \delta_{zc}^2 \end{bmatrix} \quad Q^r = \begin{bmatrix} \delta_{xr}^2 & & \\ & \delta_{yr}^2 & \\ & & \delta_{zr}^2 \end{bmatrix}$$

(17)

where $(\delta_{xc}, \delta_{yc}, \delta_{zc})$, $(\delta_{xr}, \delta_{yr}, \delta_{zr})$ are three components of the typical accuracy values the photogrammetry system and of the robot. Substituting Q^c and Q^r into Equations (15) and (16), we can obtain the optimum fused result.

The covariance of X^i is given to be:

$$V^i(X) = J^i Q^i J^{iT}$$

(18)

The fused covariance of X is given below:

$$V(X) = \left\{ \sum_{i=1}^{n} (J^i Q^i J^{iT})^{-1} \right\}^{-1}$$

(19)

It obviously can be deduced from Equation (19) that $V(X) < V^i(X)$; then $V(X) < V^i(X)$. This proves that the fusion accuracy is better than the local accuracy.

3. EXPERIMENTS AND DISCUSSION

To improve the accuracy of a robot manipulator, there are two steps in this paper. Firstly, the accuracy of a robot manipulator can be improved through the calibration for the kinematics parameters of the robot by the photogrammetry system. In addition, through calibrating the kinematics parameters of the robot, we can obtain a transformation relationship between the coordinate system of the photogrammetry system and robot. This makes the base coordinate system of the robot be a local unified coordinate system. Secondly, using the pose data of the calibrated robot and the online measurement of the multi-sensor combined measurement system (MCMS), three kinds of measurement data can be obtained. They are from the calibrated robot, photogrammetry system and inclinometer. The result can be improved by fusing these redundant data through KF and MOIFA. Therefore, four experiments are designed in this paper. Firstly, as an important part in MCMS, since the photogrammetry system directly monitors the robot pose, the accuracy of the photogrammetry system is crucial to the whole measurement system. The measurement errors of the photogrammetry system often appear to be due to the distortion on the edge of the FOV and the mistaken identity to the target. Therefore, it is necessary to research the repeatability accuracy of the photogrammetry system in its FOV, which is detailed in Section 3.1. Secondly, to improve the pose accuracy of the robot, the primary work is the calibration of the robot. Therefore, the experiment using the photogrammetry system to calibrate the robot is designed in Section 3.2. Thirdly, to observe the effect of the two data fusion methods, we design a simulation in Section 3.3. At last, in Section 3.4, a lab experiment is designed for verifying the result of the simulation. In this paper, all accuracy values (simulations and measurements) are given through the "three-sigma rule", which is a method of eliminating the gross error by thrice the standard error in the theory of errors.

3.1. Repeatability Precision of Photogrammetry System

One of the ways to improve the photogrammetry system accuracy is to search for the optimal range of the FOV. As shown in Figure 2a, in order to test the accuracy of the photogrammetry system in the FOV, an experiment is designed as follows. The space of $1 \times 0.8 \times 1$ m is divided into five planes from bottom to top, each of which contains 8~9 points. Then, 9~12 lines are formed through connecting the adjacent points. We move the robot manipulator to these points sequentially, and the photogrammetry system measures the coordinates of each point five times. In this paper, the laser tracker offered by FARO Co., Ltd. measures all points three times, and the result is used as the reference value. The FARO Xi laser tracker in the lab, whose uncertainty of the absolute distance meter (ADM) is 10 μm + 1.1 μm/mL, has been verified by the National Metrology Institute of China (NIM CDjx2008-0782). The measurement of the lines is similar as the points. The image of the experimental field is shown in Figure 2b.

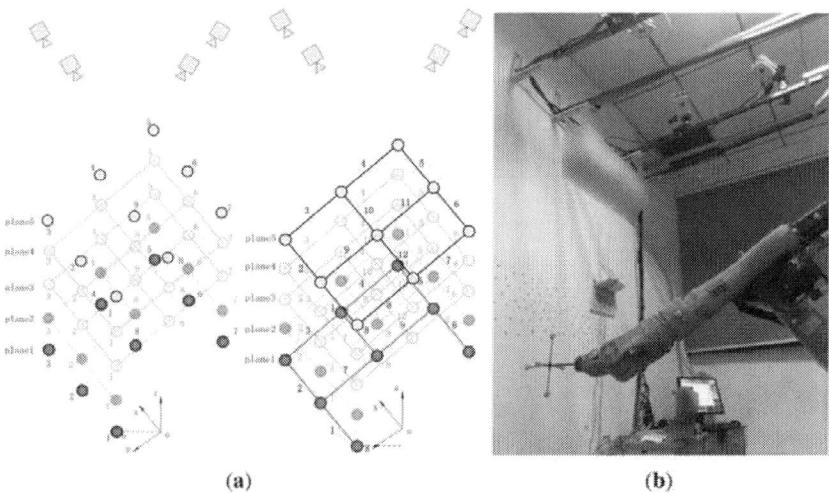

(a)   (b)

Figure 2. (a) Experimental principle of repeatability precision of the photogrammetry system; (b) the image of the experimental field.

3.1.1. Results of the Repeatability Precision of the Photogrammetry System

As shown in Table 1, the standard deviations of the position for 43 points are calculated by five groups of data, where $\delta_x, \delta_y, \delta_z$ are the standard deviations of three dimensions. Due to space limitations, we extract the data of Planes 1 and 5. δ_d is the compound standard deviation of $\delta_x, \delta_y, \delta_z$.

Table 1. The standard deviations of the repeated measurement of points (units: mm).

Plane 1	δ_x	δ_y	δ_z	δ_d
1	0.046	0.929	0.303	0.979
2	0.119	0.872	0.246	0.914
3	0.128	1.058	0.242	1.093
4	0.218	0.800	0.325	0.891
5	0.106	1.189	0.229	1.215
6	0.075	1.028	0.275	1.067
7	0.064	0.972	0.283	1.015
8	0.231	1.126	0.272	1.181
Plane 5	δ_x	δ_y	δ_z	δ_d
35	13.902	7.396	10.661	19.017
36	0.062	0.950	0.255	0.985
37	4.709	9.126	1.822	10.431
38	0.075	1.057	0.256	1.090
39	10.487	9.584	25.830	29.479
40	0.095	0.920	0.286	0.968
41	0.102	0.896	0.406	0.989
42	0.064	0.989	0.285	1.031
43	0.068	1.007	0.273	1.046

In order to show the distribution of the repeatability error in the FOV of the photogrammetry system, the histograms of the standard deviation are drawn in Figure 3. Figure 3a–e shows the error of Planes 1~5, and Figure 3f shows the merged errors of all planes.

Some phenomena are observed in Figure 3. Firstly, the maximum error appears at the corner of each plane, such as Points 1, 3, 5, 7. Their merged errors are much larger than the other points, as shown in Figure 3. Except for these cornered points, the errors of the rest of the points are almost similar. Secondly, with the decreasing of the distance between the plane and the camera, the values and the number of the errors increase. Thirdly, the errors in the direction of x and z are smaller than y.

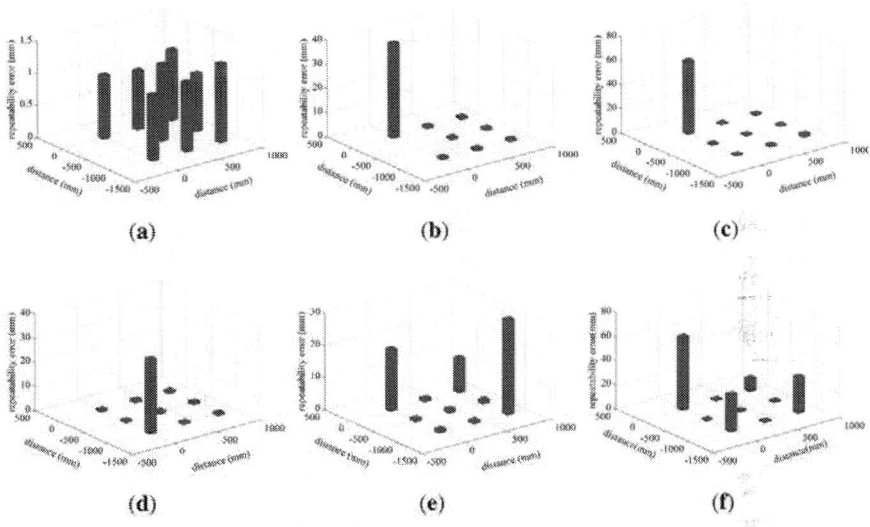

Figure 3. The histograms of the error of five planes and the merged error of all planes.

In this paper, TENYOUNfull body motion capture 3DMoCap-GC130 is used as a visual sensor. It takes more than 6 mm of prime lens as the optical lens of the camera. Its measurement range is more than $1 \times 1 \times 1$ m. As we known, the FOV of the camera will enlarge with increasing of the photograph distance. The size of Plane 5 is $1 \times 0.8 \times 1$ m, which can be considered to be the range close to the limitation of the measurement range. Plane 1, the length size of which is $1 \times 0.8 \times 2$ m, is in a reasonable measurement range. Therefore, the error of Plane 5 is the largest in all planes and that of Plane 1 is the smallest. The result in Table 1 and Figure 3 shows that the repeatability error in the range of $1 \times 0.8 \times 1 \sim 2 \times 0.8 \times 1$ m has higher accuracy. The best accuracy is in the center of the range, the average values of which are δ_x = 0.124 mm, δ_y = 0.997 mm, δ_z = 0.272 mm, δ_d = 1.045 mm.

Secondly, we discuss the situation of the lines. As shown in Table 2, the standard deviations of 54 lines are calculated in comparison to the data of the laser tracker. Similarly, we select the data of Plane 1 and Plane 5. d_l is the length measured by the laser tracker, and d_c is the length measured by the photogrammetry system. Δ_d is the difference of the laser tracker and the photogrammetry system. The errors of the lines of each plane are drawn in Figure 4.

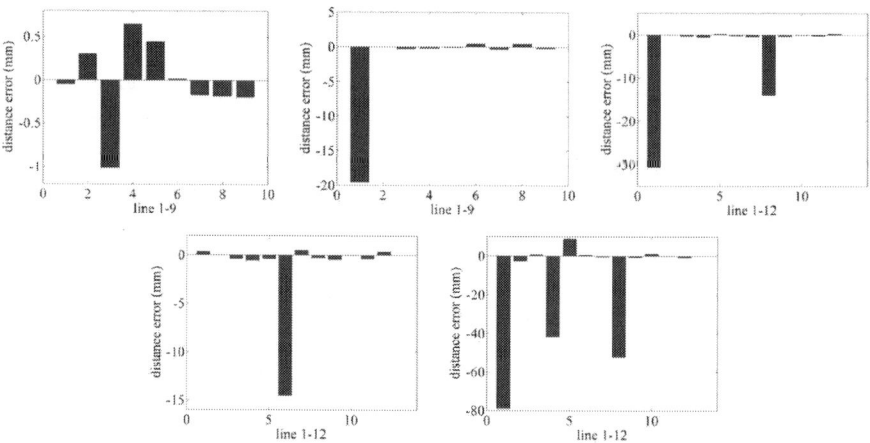

Figure 4. The error of the lines of five planes.

From Table 2 and Figure 4, we can find that the error of the lines has a similar phenomena as the points. The error of the lines increases with the decreasing of the distance between the plane and camera. The peak of the error appears in Lines 3, 4, 6, 8, which connect to the corner Points 1, 3, 5, 7. Accordingly, a conclusion can be drawn that the accuracy of the photogrammetry system is much higher in the range of 1 × 0.8 × 1 ~ 2 × 0.8 × 1 m. The best accuracy of the lines is located in the center of the FOV, and its average value is Δd = 0.478 mm.

Table 2. The standard deviations of lines (units: mm).

Plane 1	d_l	d_c	Δd
1	400.202	400.249	−0.047
2	400.482	400.179	0.302
3	500.253	501.264	−1.010
4	500.140	499.492	0.647
5	400.794	400.350	0.443
6	400.509	400.491	0.018
7	500.325	500.498	−0.172
8	400.399	400.583	−0.184
9	500.112	500.306	−0.193

Plane 5	d_l	d_c	Δd
1	321.442	400.547	−79.104
2	398.357	401.150	−2.793
3	501.106	500.544	0.562
4	458.928	500.737	−41.809
5	409.822	401.110	8.711
6	401.108	400.693	0.415
7	499.526	500.038	−0.511
8	447.228	499.478	−52.249
9	499.533	500.300	−0.767
10	401.440	400.309	1.131
11	499.971	500.153	−0.181
12	401.508	402.316	−0.808

Therefore, the photogrammetry system can be used as the calibration instrument. The measured data also can be used as the feedback data to compensate for the errors of the robot in its effective measurement range.

3.2. Calibrating Method for the Robot

One of the ways to improve the pose accuracy for the robot manipulator is to calibrate the kinematics parameters of the robot. The photogrammetry system is used to monitor the pose of robot, so that it is reasonable to calibrate the robot by means of it. This paper proposes a calibrating method for the position error of the robot that is based on the D-H model. The first step of this method is to build a model of the coordinate transformation between the coordinate system of the photogrammetry system and the robot. Then, a kinematics parameter model of the robot manipulator is established according to the differential equation of the kinematics parameters error for the robot axes. The measured data of the photogrammetry system are converted into the coordinate system of the robot. Through comparing with the converted measured data, the parameters of the robot, such as the robot kinematics parameter, target installation error and transferred error of coordinate system, can be calibrated. A simple description of the main principle is shown below. The details of the calibrated method are shown in [17,18].

Figure 5 shows a simple model of the robot calibration. $O_pX_pY_pZ_p$ is the coordinate system of the photogrammetry system. $O_oX_oY_oZ_o$ is the actual base coordinate system of the robot. $O_rX_rY_rZ_r$ is the virtual base coordinate system of the robot measured by the photogrammetry system. The difference between the $O_oX_oY_oZ_o$ and $O_rX_rY_rZ_r$ is caused by the errors of the transfer matrix.

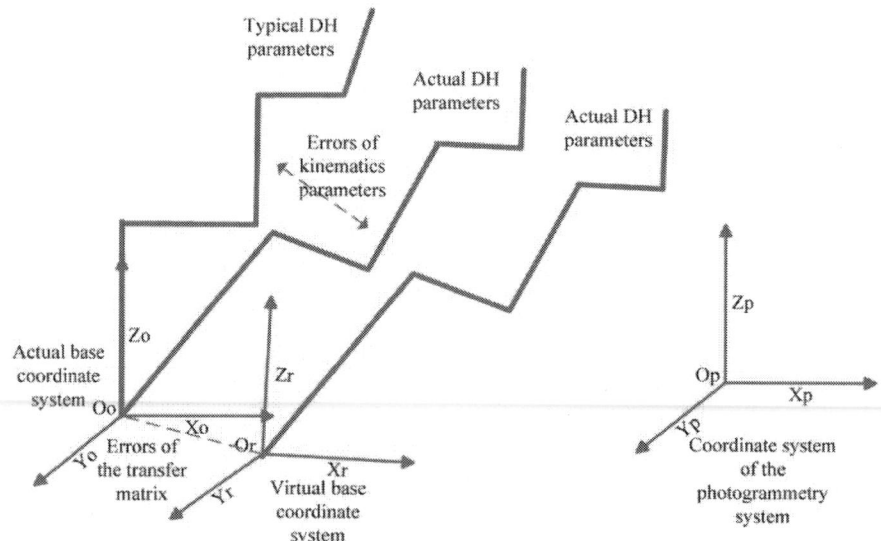

Figure 5. Simplified model of the robot calibration.

In order to obtain the position error of the robot, we must unify the coordinate systems of the photogrammetry system to the robot firstly.

Assume
$$T_p^r = \begin{bmatrix} r_{11} & r_{12} & r_{13} & x \\ r_{21} & r_{22} & r_{23} & y \\ r_{31} & r_{32} & r_{33} & z \\ 0 & 0 & 0 & 1 \end{bmatrix}$$
is the transfer matrix from the coordinate system of the photogrammetry system $O_p X_p Y_p Z_p$ to the virtual base coordinate system of the robot $O_r X_r Y_r Z_r$. Rotating Axis 1 of the robot with a certain degree, the photogrammetry system can obtain a group of data. Fitting the data, we obtain the vector of the direction z, which is the third component of Trp. Similarly, the vector of the direction x, i.e., the first component of Trp, can be obtained by rotating Axis 2 of the robot. Then, the vector of the direction y, i.e., the second component of Trp, can be calculated by the cross-product of vector z and x. The translation vector (x, y, z) also can be fitted by the data.

Suppose that the error model of the transfer matrix from the $O_r X_r Y_r Z_r$ and $O_o X_o Y_o Z_o$ is expressed as Equation (20):

$$T_r^o = \begin{bmatrix} 1 & -\delta_z & \delta_y & d_x \\ \delta_z & 1 & -\delta_x & d_y \\ -\delta_y & \delta_x & 1 & d_z \\ 0 & 0 & 0 & 1 \end{bmatrix}$$

(20)

where δ_x, δ_y, δ_z are the errors of the rotation matrix and d_x, d_y, d_z are the errors of the translation matrix.

In addition, the cooperation target of the photogrammetry system, which is set up at the end axis of the robot, should be considered as an additional axis, Axis 7. Therefore, the transfer matrix from Axis 6 to Axis 7 is shown in Equation (21).

$$T_6^7 = \begin{bmatrix} 1 & 0 & 0 & t_x \\ 0 & 1 & 0 & t_y \\ 0 & 0 & 1 & t_z \\ 0 & 0 & 0 & 1 \end{bmatrix}$$

(21)

where t_x, t_y, t_z are the translation vectors, which can be measured previously.

Therefore, the transfer matrix from the coordinate system of the photogrammetry system to the coordinate system of the robot manipulator is shown in Equation (22).

$$T = \left(\sum_{i=1}^{7} T_{i-1}^i \right) T_r^o T_p^r$$

(22)

Assume $B_p = \begin{bmatrix} r_{1p} & r_{2p} & r_{3p} & p_{xp} \\ r_{4p} & r_{5p} & r_{6p} & p_{yp} \\ r_{7p} & r_{8p} & r_{9p} & p_{zp} \\ 0 & 0 & 0 & 1 \end{bmatrix}$ is the pose of a certain point in the coordinate system of the photogrammetry system, where $r_{1p} \sim r_{9p}$ are the attitude parameters and $p_{xp} \sim p_{zp}$ are the position parameters. The converted pose of this point from the coordinate system of the photogrammetry system to the coordinate system of the robot manipulator $B_r = \begin{bmatrix} r_{1r} & r_{2r} & r_{3r} & p_{xr} \\ r_{4r} & r_{5r} & r_{6r} & p_{yr} \\ r_{7r} & r_{8r} & r_{9r} & p_{zr} \\ 0 & 0 & 0 & 1 \end{bmatrix}$ can be obtained by calculation of $B_r = T \cdot B_p$ using Equation (22). Then, the Z-Y-Z Euler angles (ϕ, θ, ψ), which express the attitude angles of the robot manipulator, can be obtained as shown in Equation (23).

$$\begin{cases} \phi = \arctan \dfrac{r_{6r}}{r_{3r}} \\[2mm] \theta = \arctan \dfrac{\sqrt{r_{7r}^2 + r_{8r}^2}}{r_{9r}} \\[2mm] \psi = \arctan \dfrac{r_{8r}}{-r_{1r}} \end{cases}$$

$$(23)$$

The link parameters of the robot are the most significant impact factors on the position error of the robot. In the D-H model, they are the length of the link a, the displacement of the link d, the angle of rotation α and the link angle of each axis θ. In this paper, we adopt a series robot of six axes, so we can totally obtain 24 kinematics parameters $\Delta a_{1\sim4}, \Delta d_{1\sim4}, \Delta \alpha_{1\sim4}, \Delta \theta_{1\sim4}$. In addition, the transfer matrix from the coordinate system of the robot tool-center-point (TCP) to the end axis of the robot has nine rotation and translation error variables, as Equation (20) shows. Therefore, there are 33 parameters of the robot that need to be calibrated.

According to the distance error model of a series robot [17], the relationship of the distance error and position error is shown as Equation (24).

$$\Delta l(i, i+1) = \left[\frac{x_R(i+1) - x_R(i)}{l_R(i, i+1)}, \frac{y_R(i+1) - y_R(i)}{l_R(i, i+1)}, \frac{z_R(i+1) - z_R(i)}{l_R(i, i+1)} \right] \cdot (\overrightarrow{dp_{i+1}} - \overrightarrow{dp_i})$$

$$(24)$$

where i is the number of the point on the command trajectory. $l_R(i, i+1)$ is the distance between the point i and i + 1 on the command trajectory. (x_R, y_R, z_R) is the position coordinate components of a certain point in the robot coordinate system $O_oX_oY_oZ_o$. Δl is the distance error, i.e., the difference value between the theoretical position and practical position. In this paper, the theoretical value is obtained by the robot encoder, and the practical value is obtained by the photogrammetry system. dp is the vector for the position error of the robot.

Because of the impact of four link parameters of the robot, it will cause the position error for the adjacent axes of the robot dTii−1, which can be expressed as Equation (25).

$$dT^i_{i-1} = \frac{\partial T^i_{i-1}}{\partial \theta_i} \Delta \theta_i + \frac{\partial T^i_{i-1}}{\partial \alpha_i} \Delta \alpha_i + \frac{\partial T^i_{i-1}}{\partial a_i} \Delta a_i + \frac{\partial T^i_{i-1}}{\partial d_i} \Delta d_i$$

(25)

If each of the two adjacent axes are influenced by the link parameters, the transformation from the base coordinate system of the robot to the coordinate system of the robot manipulator can be expressed as Equation (26). In this paper, N = 6.

$$T^N_0 + dT^N_0 = \prod^N_{i=1} (T^i_{i-1} + dT^i_{i-1}) = \prod^N_{i=1} (T^i_{i-1} + T^i_{i-1} \Delta_i)$$

(26)

where $\Delta_i = T_{\theta i} \cdot \Delta \theta_i + T_{\alpha i} \cdot \Delta \alpha_i + T_{ai} \cdot \Delta a_i + T_{di} \cdot \Delta d_i$. Additionally, $T_{\theta i}, T_{\alpha i}, T_{ai}, T_{di}$ can be obtained by the calculation of the robot kinematics parameters. Through expanding Equation (26) with a large number of simplifications and combinations, the position error of the robot manipulator can be obtained as given in Equation (27). More detail about the calculation procedures can be found in [17].

$$\Delta p = [d_{tx} d_{ty} d_{tz}]^T$$
$$= \begin{bmatrix} k^x_{1\theta} & k^x_{1\alpha} & k^x_{1a} & k^x_{1d} & k^x_{2\theta} & \cdots & k^x_{6\theta} & k^x_{tx} & k^x_{ty} & k^x_{tz} \\ k^y_{1\theta} & k^y_{1\alpha} & k^y_{1a} & k^y_{1d} & k^y_{2\theta} & \cdots & k^y_{6\theta} & k^y_{tx} & k^y_{ty} & k^y_{tz} \\ k^z_{1\theta} & k^z_{1\alpha} & k^z_{1a} & k^z_{1d} & k^z_{2\theta} & \cdots & k^z_{6\theta} & k^z_{tx} & k^z_{ty} & k^z_{tz} \end{bmatrix}$$
$$[\Delta \theta_1 \Delta \alpha_1 \Delta a_1 \Delta d_1 \Delta \theta_2 \cdots \Delta d_6 \Delta t_x \Delta t_y \Delta t_z]^T$$
$$= B_i \Delta q_i$$

(27)

where Δp is the position error of the robot manipulator. (d_{tx}, d_{ty}, d_{tz}) are the Cartesian coordinate components of the position error.

$$B_i = \begin{bmatrix} k_{1\theta}^x & \cdots & k_{tz}^x \\ \vdots & \ddots & \vdots \\ k_{1\theta}^z & \cdots & k_{tz}^z \end{bmatrix}$$ is the parameter matrix related to the typical

position value of the robot manipulator. $\Delta q_i = [\Delta \theta_1 \cdots \Delta t_z]^T$ are the kinematics parameters of six-degree series robot and the translation error parameters from the coordinate system of the robot TCP to the end axis of the robot. In Equation (27), the left side is the position error at each point measured by the photogrammetry system. The right side is the kinematics errors that need to be corrected. These errors can be revised by the least squares method in the generalized inverse matrix sense.

After calibrating by the photogrammetry system, the position error of the robot can be less than 1 mm. Figure 6 shows the position error of 71 points.

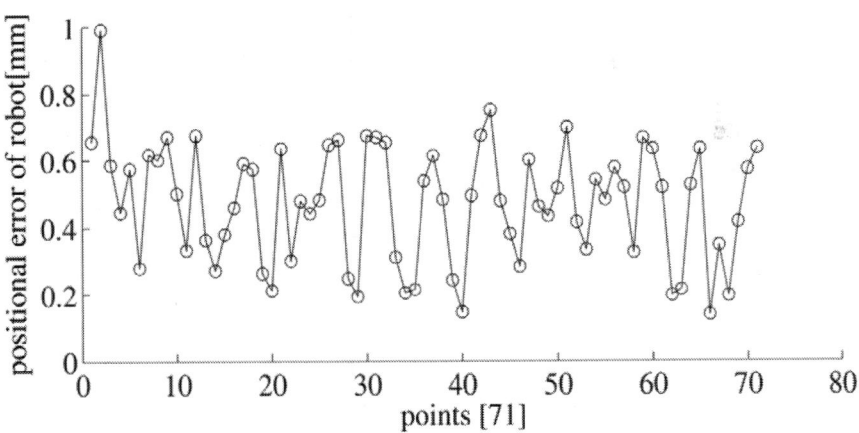

Figure 6. The position error of the robot after calibration.

3.3. Simulation Test of the Sensor Data Fusion Methods

In terms of the position of the robot manipulator, there are two kinds of measurement data: one is obtained from the photogrammetry system and the other one is from the encoder of the robot. We propose two sensor data fusion methods to fuse the two kinds of position data. In order to compare the two methods, a simulation test is developed in MATLAB. One-hundred random points are created in a space of $100 \times 100 \times 100$ mm to simulate the actual positions of the robot. For the purpose of simulating the actual value, each point is mixed with an error. It follows the normal distribution, which is determined by the typical value δ of the measured instruments. In this paper, the typical accuracy values of the photogrammetry system are $\delta_{xc} = \delta_{yc} = \delta_{zc} = 0.15$ mm.The typical accuracy values of the robot are δ_{xr} = 0.157 mm, δ_{yr} = 0.087 mm, δ_{zr} = 0.043 mm. The typical values are calibrated by the FARO Xi laser tracker. Then, 100 points are fused by using KF and MOIFA.

In the first method, the simulated measured value of the photogrammetry system and robot, $Y^1(k)$ and $Y^2(k)$, are input into KF, as described in Equations (1)–(5), to obtain the estimated state variables, $X^1(k + 1)$ and $X^2(k + 1)$. Then, they are fused using the weight matrix W described in Equation (14), which is determined by the Jacobian matrices and covariance matrices of the photogrammetry system and robot. The fused error using KF is drawn in Figure 7a. In the second method, the simulated value of the photogrammetry and robot are fused as described in Equations (15)–(17). The fusion errors of MOIFA are drawn in Figure 7b.

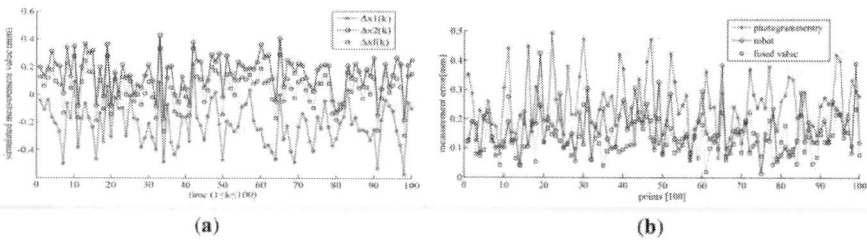

(a)   (b)

Figure 7. (a) The simulated fused error using the KF method; (b) The simulated fused error using the multi-sensor optimal information fusion algorithm (MOIFA).

3.3.1. Results of the Simulation Test of the Data Fusion Methods

As shown in Table 3, Δx_1, Δx_2 are the estimated errors of the photogrammetry and robot after fusing by the KF method, respectively. Δx_f is the fused error. Δ_{CM}, Δ_{RB} are the errors of the photogrammetry and robot, respectively. Δ_f is the fused error after fusing by MOIFA.

Table 3. Simulation results of the data fused by KF and MOIFA (units: mm).

Δx_1	Δx_2	Δx_f
−0.198	0.139	0.043

Δ_{CM}	Δ_{RB}	Δ_f
0.240	0.156	0.129

Since the photogrammetry system has a bigger typical accuracy value, the error of the photogrammetry system is bigger than that of the robot, as shown in Table 3. It is indicated in Figure 7a,b and Table 3 that either of the methods can reduce the error after the data fusion. Through calculating the data in Table 3, the error is reduced by 78.2% with KF and by 46.1% with MOIFA. As shown in Table 3, KF has smaller fused errors than MOIFA. In addition, KF can predict the state variables of next moment, which is suitable to be applied in the dynamic measurement and compensation. MOIFA can only analyze the ready-measured data. However, it will cause the hysteresis of the real-time compensation. It should be noted that the measuring range is 0~100 mm in this experiment. KF is a linear filter, so that its fused error will be enlarged with increasing the measuring range. This will be verified in the lab experiment in Section 3.4.

3.4. Verified Experiment in the Lab

In order to validate the conclusion of the simulation test, a data fusion-verified experiment in lab is designed. For the purpose of measuring the pose of the robot manipulator, a five-ball target frame of the photogrammetry system and the inclinometer are set up at the end of the robot through a special fixture, as shown in Figure 1. Seventy six points are located on the surface of a 200-mm radius sphere, which is in the space of $1 \times 1 \times 1$ m in the front of the robot. Obviously, these points must also be located in the effective measurement area of the photogrammetry system, as described in Section 3.1. For acquiring the stable data, the robot stays in each position for 7 s to offer enough measurement time for the photogrammetry system. Since the robot has been calibrated in Section 3.2, in this experiment, all data are converted into the base coordinate system of the robot. The measurement value of the FARO Xi laser tracker is used as the reference value. Then, 71 picked points are fused using KF and MOIFA.

The same as the process of the simulate experiment described in Section 3.3, the fused error using KF is drawn in Figure 8a, and the fused error using MOIFA is drawn in Figure 8b.

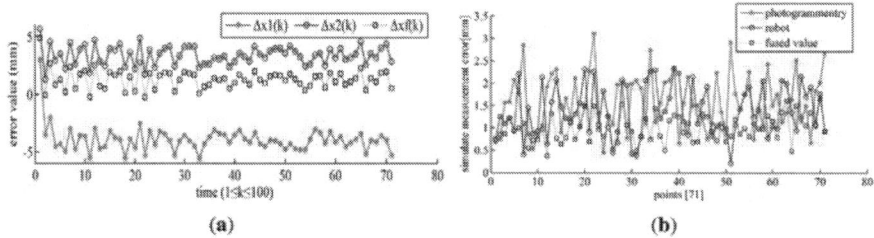

Figure 8. (a) The experimental error using KF; (b) the experimental error using MOIFA.

3.4.1. Result of the Verified Experiment in the Lab

The average values of the measurement error are shown in Table 4.

Table 4. Experimental results of the data fused by KF and MOIFA (units: mm).

Δx_1	Δx_2	Δx_f
−3.940	3.379	1.297
Δ_{CM}	Δ_{RB}	Δ_f
1.587	1.386	0.981

It is indicated in Table 4 that in the lab experiment, the errors of the photogrammetry system are bigger than that of the robot. This is because its typical value is bigger than the robot, that same as the simulation experiment shown in Section 3.3.1. The error of the photogrammetry system Δ_{CM} in the lab experiment is smaller than the simulation test. This illustrates that the accuracy of the photogrammetry system in its effective FOV is improved, as we had tested in the repeatability experiment in Section 3.1.1. It is seen in Figure 8a,b and Table 4 that both of the fused methods can reduce the error after the data fusion. Through the data calculation in Table 4, the error is reduced by 67.3% with KF and by 38.2% with MOIFA. This is a little smaller than the results of the simulation, but both of them have common trends. However, the value of the fused error using KF is a little larger than the error of the simulation. The reason is that KF is a linear filter, which had been discussed previously in Section 3.3.1. Comparing to the other works, Yauheni and Jerzy of Warsaw University of Technology obtain the improved value of the positioning accuracy for the robot end-effector $\Delta L = 2.39$ mm using the method of joint error mutual compensation. Li Junmin and Wang Jinge et al. improved the pose accuracy of the robot to 2.2 mm based on the unit quaternion and the prediction of the pose estimation accuracy. It is indicated that the multiple-sensor combination measuring system (MCMS) proposed in this paper has good performance for improving the pose accuracy of the robot.

Therefore, a conclusion can be made in comparison to the simulation that both of the data fusion methods can lead to the improvement of the results. MOIFA has more stable accuracy of the fusion, but it has no predicted function, which would cause hysteresis in the feedback control system of the robot. KF is widely applied in the areas of robotics and aviation. It possesses the predicted function of the next moment, which is suitable for real-time measurement and compensation. However, its predicted error would be enlarged with increasing measurement range, since it is a kind of linear filter. According to the features of the two data fusion methods, we can adopt KF to fuse data when dynamic and real-time measurement is needed, as well as when the measurement range is small. Otherwise, MOIFA can be adopted when the static and offline measurement is needed, as well as when the measurement range is large.

4. CONCLUSIONS

In this paper, we proposed a multi-sensor combination measuring system (MCMS) and two sensor data fusion methods to improve the pose accuracy of industrial robots. The advantage of this method is that it is automatic and does not involve environmental intervention. To ensure the accuracy of the measured sensor, this paper researched the repeatability precision of the photogrammetry system and the robot calibration by means of the photogrammetry system. The experimental results show that the best accuracy of the photogrammetry system is in the center of the FOV, which is the range of $1 \times 0.8 \times 1 \sim 2 \times 0.8 \times 1$ m. The position error of the robot manipulator is less than 1 mm after being calibrated by the photogrammetry system. In order to improve the accuracy of the robot pose, we propose two kinds of data fusion methods to fuse the redundant information gathered from the multiple sensors. Through comparing with the simulation and lab experimental results, KF possesses a predicted function of the next moment that is suitable for real-time measurement and compensation. However, its predicted error would be enlarged with an increasing measurement range, since it is a kind of linear filter. On the other hand, MOIFA possesses the stable accuracy of fusion, but it is not capable of the predicting function. This will cause hysteresis in the feedback control system of the robot. Therefore, both of the methods can reduce the pose error of the robot by 38%~78%. The choice of method is dependent on the requirements of the measurement. The experimental and theoretical results provided the basis for an industrial application of the robot pose measurement and compensation. Future works will include the real-time transferring of data, online control and compensation for the pose of the robot manipulator.

ACKNOWLEDGEMENTS

This research was supported by the Natural Science Foundation of China (NSFC) No. 51275350 and the Specialized Research Fund for the Doctoral Program of Higher Education of China No. 20110032110045.

AUTHOR CONTRIBUTIONS

Bailing Liu, Fuming Zhang and Xinghua Qu conceived and designed the experiments; Bailing Liu performed the experiments; Bailing Liu and Fumin Zhang analyzed the data; Xinghua Qu contributed analysis tools; Bailing Liu wrote the paper.

REFERENCES

1. Yauheni, V.; Jerzy, K. Application of joint error mutual compensation for robot end-effector pose accuracy improvement. J. Intell. Robot. Syst. Theory Appl. **2003**, 36, 315–329.

2. Jian, Y.Z.; Chen, Z.; Da, W.Z. Pose accuracy analysis of robot manipulators based on kinematics. Adv. Manuf. Syst. **2011**, 201–203, 1867–1872.

3. Du, G.; Zhang, P. Online serial manipulator calibration based on multisensory process via extended kalman and particle filters. IEEE Trans. Ind. Electron. **2014**, 61, 6852–6859.

4. Frank, S.C. The Method of Recovering Robot TCP Positions in Industrial Robot Application Programs. Proceedings of the 2007 IEEE International Conference on Mechatronics and Automation, Harbin, China, 5–8 August 2007; pp. 805–810.

5. Hans, D.R.; Beno, B. Visual-model-based, real-time 3D pose tracking for autonomous navigation: Methodology and experiments. Auton. Robot **2008**, 25, 267–286.

6. Kaijen, H.L.; Pack, K.; Tomás, L.P. Robust grasping under object pose uncertainty. Auton. Robot **2011**, 201–203, 253–268.

7. Qu, W.W.; Dong, H.Y.; Ke, Y.L. Pose accuracy compensation technology in robot-aided aircraft assembly drilling process. Acta Aeronaut. Astronaut. Sin. **2011**, 32, 1951–1960.

8. Camille, S.C.; Rainer, S.; Frank, B.; Franck, S.M. Registration of arbitrary multi-view 3D acquisitions. Comput. Ind. **2013**, 64, 1082–1089.

9. Thomas, L. Close range photogrammetry for industrial applications. ISPRS J. Photogramm. Remote Sens. **2010**, 65, 558–569.

10. Ba, H.X.; Zhao, Z.G. A survey on mathematic models and methods of multisensor data fusion. Ship Sci. Technol. **2005**, 27, 48–53.

11. Kalman, R.E. A new approach to linear filtering and prediction problems. J. Basic Eng. **1960**, 82, 35–46.

12. Abidi, M.A.; Gonzalez, R.C. Data Fusion in Robotics and Machine Intelligence; Academic Press: Waltham, MA, USA, 1992; Volume 19, pp. 888–896.

13. Nakamura, Y.; Zu, Y. Geometrical Fusion Method for Multi-Sensor Robotic Systems. Proceedings of the 1989 IEEE International Conference on Robotics and Automation, Scottsdale, AZ, USA, 14–19 May 1989; Volume 2, pp. 668–673.

14. Langlois, D.; Elliott, J.; Croft, E.A. Sensor Uncertainty Management for an Encapsulated Logical Device Architecture, Part II: A Control Policy for Sensor Uncertainty. Volume 1–6, 4288–4293.

15. Nandi, G.C.; Mitra, D. Development of a Sensor fusion Strategy for Robotic Application Based on Geometric Optimization. J. Intell. Robot. Syst. **2002**, 35, 171–191.

16. John, M.; Allan, S.; Manhtriet, H.; Arnold, F. Sensor Fusion of Laser Trackers for Use in Large-Scale Precision Metrology in Intelligent Manufacturing. Proc. SPIE Intell. Manuf. **2004**, 5263.

17. Ren, Y.J. The Study on Main Body Calibration Technique of Measuring Robot; Tianjin University: Tianjin, China, 2007; pp. 71–76.

18. Li, R.; Qu, X.H. Study on calibration uncertainty of industrial robot kinematics parameters. Chin. J. Sci. Instrum. **2014**, 35, 2192–2199.

CHAPTER 8

A Mechanical Musculo-Skeletal System for a Human-Shaped Robot Arm

Koichi Koganezawa

Department of Mechanical Engineering, Tokai University, 4-1-1 Kitakaname, Hiratsuka, Kanagawa 259-1292, Japan.

ABSTRACT

This paper presents a mechanical system with a similar configuration to a human musculo-skeletal system for use in anthropomorphic robots or as artificial limbs for disabled persons. First, a mechanical module called ANLES (Actuator with Non-Linear Elasticity System) is introduced. There are two types of ANLES: the linear-type ANLES and rotary-type ANLES. They can be used as a voluntary muscle in a wide-range of musculo-skeletal structures in which at least double actuators work in an antagonistic setup via some elastic elements. Next, an application of the two types of ANLES to a two-degree-of-freedom (DOF) manipulator that has a similar configuration to the human elbow joint is shown. The experimental results of the joint stiffness and joint angle control elucidate that the developed mechanism effectively regulates joint stiffness in the same way as a musculo-skeletal system.

KEYWORDS

musculo-skeletal system; non-linear elasticity; antagonistic alignment; stiffness control

1. INTRODUCTION

Dexterous behaviors of human articulations are mostly due to the adjustability of their stiffness in response to external interferences. The musculo-skeletal system of human articulations is able to regulate stiffness prior to the actual performance of a task without using exteroceptive force or tactile feedback. The key mechanism for regulating stiffness is an antagonistic structure of the musculo-skeletal system: one agonist and its antagonist muscles counteractively drive one articulation. Simultaneous stretching/relaxing of both muscles increase/decrease the stiffness of articulations.

A non-linear elasticity of individual muscles is prerequisite for the agonist-antagonist alignment to regulate stiffness. Some amount of the rotation of an articulation requires a respectively small torque under an equilibrium state of low stretching of both muscles. On the other hand, an equilibrium state under high stretching requires a respectively large torque to provide the same rotational angle. Therefore, stiffness is regulated according to the magnitude of the stretching of both muscles. It is obvious that linear elasticity does not provide such a stiffness change because a derivative of the linear elasticity with respect to joint rotation, which is only stiffness, is constant.

A vast amount of physiological studies have elucidated that skeletal muscles have such a non-linear elasticity as that mentioned above [1,2,3,4].

Some studies for investigating the stiffness of human arms elucidate that the stiffness ellipsoid of the arm's endpoint is adjustable in its volume by stretching muscles [5], but its shape is roughly determined by the arm's posture [6].

Some studies in the field of robotics deal with the antagonistic control of joints [7,8,9,10,11,12] and have noted the importance of the non-linear characteristics of elastic elements to control the stiffness of joints [9,10,11], but there have been few papers that proposed a control method of stiffness from the practical point of view, although some theoretical approaches for stiffness control have provided valuable insight [11,13,14].

This study assumes artificial joints that are controlled by at least two actuator units having a similar elastic characteristic to human voluntary muscles, called antagonistically driven joints (ADJ).

There have been some approaches composing ADJ using linear actuators that work like muscles. The most successful approach developed so far are approaches using the McKibben-type pneumatic actuator [15,16]. Although pneumatic rubber actuators inherently have non-linear elasticity, there are some drawbacks, such as the difficulty of designing non-linear elasticity, heat sensitivity, and the large volume of the apparatus necessary for supplying compressed air. The "pleated pneumatic artificial muscle" (PPAM) [17] is an approach to overcome some problems that pneumatic actuators have had, such as a short lifespan and large hysteresis in force/contract relations.

There have been some other recent approaches to develop a non-linear elastic module used to control the stiffness of ADJ [18,19,20,21] that have presented ingenious mechanical devices to design a non-linear elasticity.

Recent ardency for developing ADJ, introduced above, suggests the importance of ADJ and the potential of practical application in the near future. Because a forthcoming "Personal Care Robot" will need to be inherently safe when it interacts with external objects, especially with the human body [22], it requires mechanical resiliency and adaptability rather than a feedback control system that artificially provides them with sensor information. Our study is one approach in the research stream mentioned above to compose ADJ.

Recently, various types of mechanisms classified as variable stiffness actuators (VSAs) have been proposed [23,24,25,26,27,28,29], of which a pioneering work, known as the MIA (mechanical impedance adjuster), has been developed by Morita and Sugano [30]. The VSA approach aims to endow robots with an intrinsically safe property for use in a human-robot interactive environment. The VSA is an actuator unit that has an adjustable elastic element between the rotary joint and rotary actuator, or has a mechanism to regulate the elasticity of the elastic element. Therefore, the VSA does not aim to compose an ADJ Also, it will be explained that the VSA developed so far is rather difficult to incorporate into a joint that has multiple rotary axes similar to a wrist joint or a hip joint because of its structural nature as a rotary actuator, although there has been a design proposition to implement the VSA into a three-degree-of-freedom (DOF) joint [31].

This paper proposes an alternative mechanism of an artificial muscle to comprise the ADJ. The mechanism has a similar concept to [20,21] in the sense of yielding a non-linear elasticity by converting the force generated by a normal linear spring into a transmission process [32,33,34]. This study aims to develop an intrinsically safe device equal to the VSA approach. In our past study, we developed a muscle-like actuator: the Actuator with a Non-Linear Elasticity System (ANLES). The basic idea of the ANLES was proposed in [32] with experimental results using a simple one-DOF joint controlled by a pair of prototypical ANLESes. A linear actuator based on the ANLES, called l-ANLES, was introduced in [33] to control a joint with multiple DOFs like a wrist joint with multiplel-ANLESes. Subsequently, a rotary-type ANLES, called r-ANLES, was introduced in [34] to control the pronation/supination of a wrist joint. However, the application presented in [33,34] was very limited in the range of joint rotation and the size of the joint. Therefore, this paper presents another application of ANLES into an upper-arm that requires a wide range of joint rotation and countermeasures for the weight itself.

In the following section, two types of ANLESes are introduced: the linear type ANLES (l-ANLES) and the rotary type ANLES (r-ANLES). Both ANLESes are used as voluntary muscles to control a joint in an antagonistic setup as a musculo-skeletal system. The third section shows an application of both types of ANLES into a two-DOF manipulator that has a configuration similar to the human elbow joint. The flexion/extension of the elbow joint is controlled by a pair of l-ANLES, while the lateral/medial rotation of the upper-arm is controlled by a pair of r-ANLES. This paper also shows a weight compensation mechanism for cancelling the weight of the forearm and the end-effector. The last section is devoted to some conclusive remarks.

2. ACTUATOR WITH NON-LINEAR ELASTICITY SYSTEM

ANLES has been developed for the control of ADJ in our studies [31,32,33], working like a voluntary muscle in a musculo-skeletal system of human articulation. Therefore, a pair of ANLESes is used to control one rotary axis. There are two types of mechanical configurations in the AN LES: the linear-type ANLES (l-ANLES) and the rotary-type ANLES (r-ANLES), as explained individually below.

2.1. Structure and Design of l-ANLES

Figure 1 shows the structural parts and the assembled appearance of l-ANLES, consisting of a large lead ball screw and the Non-Linear Elasticity Module (NLEM). The NLEM consists of a guide-shaft, a torsion-spring and a transmission cylinder, as shown in Figure 2. The torque generated by the DC-motor rotates the ball screw rod that brings about the rotation or translation of the guide shaft that covers the ball screw nut. Rotation of the guide-shaft induces twisting of the torsion-spring. The diameter of the guide shaft smoothly thins down along the rotation axis so that the torsion-spring twists around the guide shaft from the edge of the wide diameter, yielding a non-linear elasticity as described below. The l-ANLES needs to transform the rotational motion to the translational motion, and vice versa, with minimum transmission loss. We therefore placed a large lead ball screw (10 mm diameter of the rod with a 10 mm lead) into the guide-shaft as shown in Figure 1.

Figure 1. Assembled view and main parts of linear type Actuator with Non-Linear Elasticity System (l-ANLES).

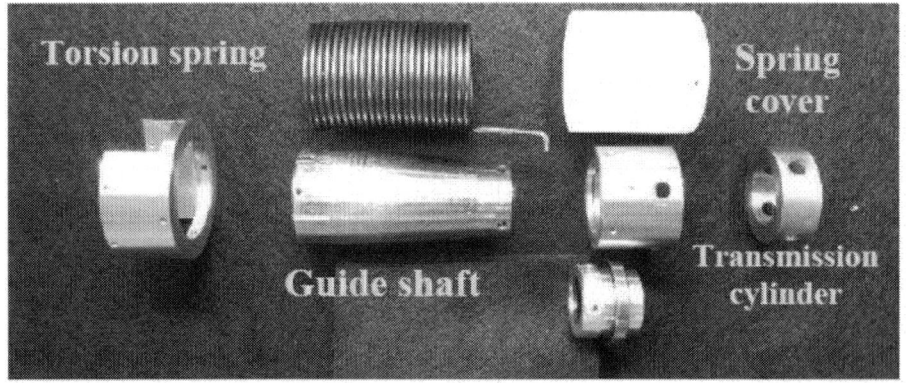

Figure 2. Non-Linear Elasticity Module (NLEM).

2.2. Structure and Design of r-ANLES

In r-ANLES, a pair of guide-shafts and torsion springs is allocated counteractively along the main shaft as shown in Figure 3. The torsion springs are twisted by the individual DC motor via the gear and wrap around the guide shaft. Therefore, it is apprehended that the r-ANLES is identical to the l-ANLES but lacked a transformation process between rotation and translation.

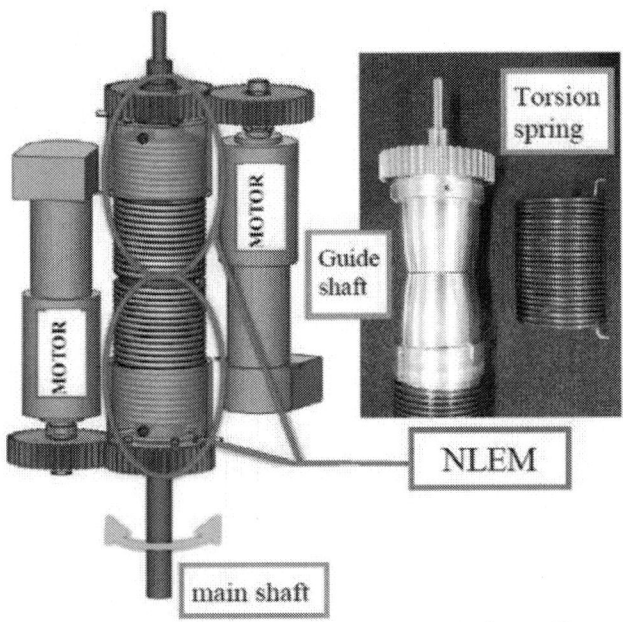

Figure 3. Rotary type ANLES (r-ANLES).

As can be easily understood in the case of the r-ANLES, two motor rotations in opposite directions rotate the main shaft (Figure 3) with no twisted springs if the motors turn at precisely the same speed. Conversely, two motor rotations in an identical direction cause the springs to twist around the guide shafts with no rotation of the main shaft while strengthening its stiffness.

The same is equally true for l-ANLES.

2.3. Design of Non-Linear Elasticity

The non-linear elasticity of *l*-ANLES and *r*-ANLES can be rigorously designed by designing the shape of the guide shaft.

$$\Delta T_a(x) = \left(EI / l_r(x)\right)\Delta\phi \tag{1}$$

where, $l_r(x)$ is the developed length of the spring wire (the length of the spring wire in part of the axial portion of $L - x$ in Figure 4) that actually works as a spring at location x. E is the modulus of the longitudinal elasticity and I is the second moment of the area of the torsion spring wire. Equation (1) leads the spring coefficient as a function of x,

$$K(x) = \Delta T_a(x) / \Delta\phi = EI / l_r(x) \tag{2}$$

Through this equation, one can obtain the relation between the torsion angle $\phi(x)$ and the torque $T_a(x)$ Hence, $T_a(x)$ and $K(x)$ may be denoted by $T_a(\phi)$ and, $K(\phi)$ respectively, in lieu of using the intermediate parameter x. Now we have a design-ability to obtain the function $T_a(\phi)$ by designing r(x), the radial function along the axis. Details of how to calculate $T_a(\phi)$ are described in Appendix A. Figure 5 shows the non-linear elasticity of the l-ANLES and the r-ANLES, which are designed according to required stiffness characteristics.

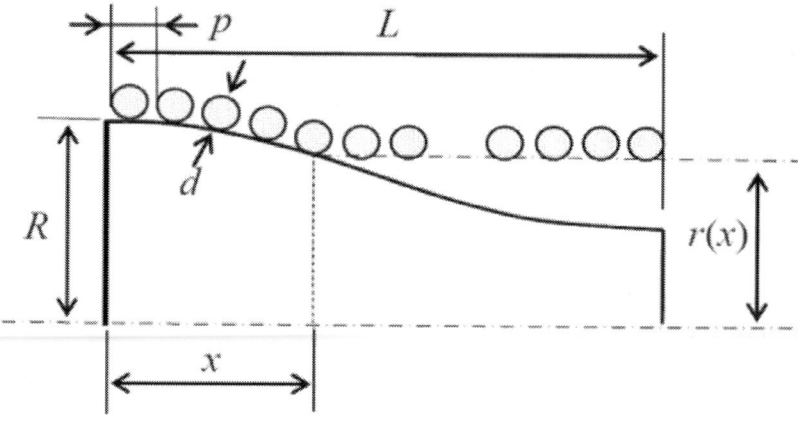

Figure 4. Model of the guide-shaft.

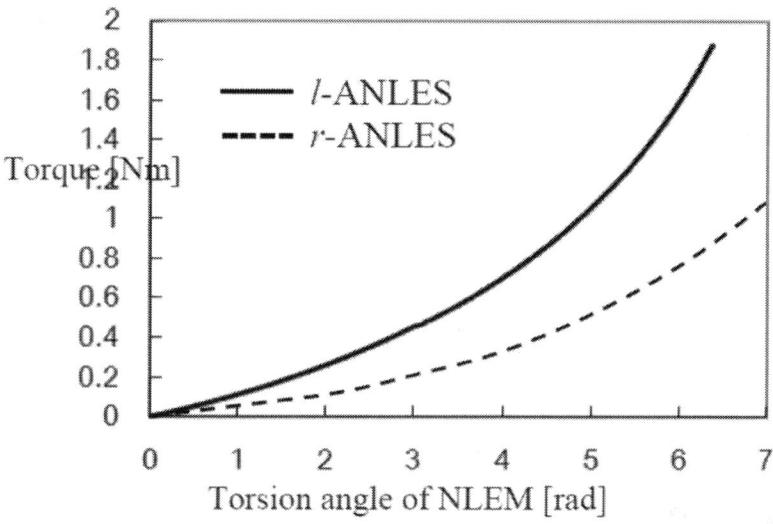

Figure 5. Non-linear elasticity of ANLES designed for elbow joint.

2.4. Stiffness of Rotary Joint

A. The case of l-ANLES

l-ANLES exerts a force vector that is also a function of the torsion angle of the spring such as,

$$\mathbf{F}_l(\phi) \equiv \left(2\pi / \ell_d\right) T_a(\phi) \mathbf{u}_l(\theta)$$

$$(3)$$

where, ℓ_d is the lead of the large lead ball screw (see Figure 1). $\mathbf{u}_l(\theta_e)$ is a unit vector along the rod axis of the l-ANLES, which depends on the joint angle θ_e. A pair of l-ANLESes exerts forces individually that counteractively affect a rotary joint axis as a torque as follows,

$$T_e = \mathbf{k}_e \cdot \left(\mathbf{r}_{mR}(\theta_e) \times \mathbf{F}_{lR}(\phi_R) - \mathbf{r}_{mL}(\theta_e) \times \mathbf{F}_{lL}(\phi_L)\right) - T_g - T_{ext}$$

$$= (2\pi / \ell_d)\mathbf{k}_e \cdot \left(T_a(\phi_R)\mathbf{r}_{mR}(\theta_e) \times \mathbf{u}_{lR}(\theta_e) - T_a(\phi_L)\mathbf{r}_{mL}(\theta_e) \times \mathbf{u}_{lL}(\theta_e)\right) - T_g - T_{ext}$$

$$(4)$$

where $\mathbf{k}_e$ is a unit vector of the rotary joint and $\mathbf{r}_{mR}(\theta_e)$ and $\mathbf{r}_{mL}(\theta_e)$ are moment arm vectors from the rotary axis to force vectors of l-ANLES, which are functions of the joint angle θ_e (see Figure 6). T_g is a torque due to a whole gravitational effect and T_{ext} is a torque generated by contact with external objects. In Equation (4) T_{ext} is not considered because the aim is to develop a robotic joint in which stiffness can be set regardless of external forces or load torques on the joint.

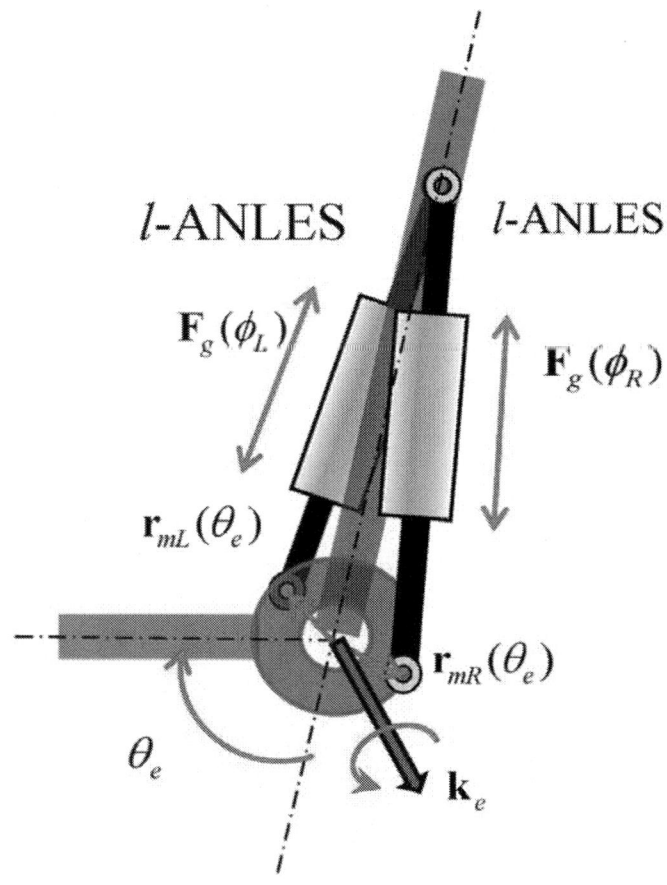

Figure 6. A rotary joint controlled by a pair of l-ANLESes.

Let us consider an equilibrium state, in which the gravitational torque is compensated by the torques due to the l-ANLES.

Denoting

$$T_a(\phi_R) \cong T_a(\bar{\phi}_R) + (\partial T_a / \partial \phi)\big|_{\phi \to \bar{\phi}_R} \Delta \phi_R , \quad T_a(\phi_L) \cong T_a(\bar{\phi}_L) + (\partial T_a / \partial \phi)\big|_{\phi \to \bar{\phi}_L} \Delta \phi_L$$

where $\bar{\phi}_R$ and $\bar{\phi}_L$ are torsion angles of the l-ANLESes provided by the individual motors and $\Delta \phi_R = \phi_R - \bar{\phi}_R$, $\Delta \phi_L = \phi_L - \bar{\phi}_L$

Assuming,

$$(2\pi / \ell_d)\mathbf{k}_e \cdot \left((\partial T_a / \partial \phi)\big|_{\phi \to \bar{\phi}_R} \Delta \phi_R \mathbf{r}_{mR}(\theta) \times \mathbf{u}_{lR}(\theta) - (\partial T_a / \partial \phi)\big|_{\phi \to \bar{\phi}_L} \Delta \phi_L \mathbf{r}_{mL}(\theta) \times \mathbf{u}_{lL}(\theta) \right) - T_g = 0$$

$$(5)$$

an equilibrium state is produced as

$$T_e = (2\pi / \ell_d)\mathbf{k}_e \cdot \left(T_a(\bar{\phi}_R)\mathbf{r}_{mR}(\theta_e) \times \mathbf{u}_{lR}(\theta_e) - T_a(\bar{\phi}_L)\mathbf{r}_{mL}(\theta_e) \times \mathbf{u}_{lL}(\theta_e) \right)$$
$$= k_R(\theta_e)T_a(\bar{\phi}_R) - k_L(\theta_e)T_a(\bar{\phi}_L) = 0$$

$$(6)$$

where $k_i(\theta_e) = (2\pi / \ell_d)\mathbf{k}_e \cdot \mathbf{r}_{mi}(\theta_e) \times \mathbf{u}_{li}(\theta_e), \quad (i = R, \text{ or } L)$

The stiffness of the rotary joint is defined as,

$$S_e \equiv \partial T_e / \partial \theta_e \, [\text{Nm/rad}]$$

$$(7)$$

If we consider changing the joint stiffness under a constant joint angle $\bar{\theta}_e$, $k_R(\bar{\theta}_e)$ and $k_L(\bar{\theta}_e)$ become constants. Therefore,

$$S_e = \frac{\partial\left(k_R(\bar{\theta}_e)T_a(\bar{\phi}_R) - k_L(\bar{\theta}_e)T_a(\bar{\phi}_L)\right)}{\partial\theta_e}$$

$$= k_R(\bar{\theta}_e)\frac{\partial T_a(\phi)}{\partial\phi}\bigg|_{\phi\to\bar{\phi}_R}\frac{\partial\phi}{\partial\theta_e}\bigg|_{\theta_e\to\bar{\theta}_e} - k_L(\bar{\theta}_e)\frac{\partial T_a(\phi)}{\partial\phi}\bigg|_{\phi\to\bar{\phi}_L}\frac{\partial\phi}{\partial\theta_e}\bigg|_{\theta_e\to\bar{\theta}_e}$$

$$\tag{8}$$

Since we can expect $k_R(\theta_e)\dfrac{\partial\phi}{\partial\theta_e}\bigg|_{\theta_e\to\bar{\theta}_e}$ and $k_L(\theta_e)\dfrac{\partial\phi}{\partial\theta_e}\bigg|_{\theta_e\to\bar{\theta}_e}$ to always have opposite signs in a counteractive configuration, as shown in Figure 6, we can calculate the absolute value of S_e as,

$$|S_e| = \frac{\partial T_a(\phi)}{\partial\phi}\bigg|_{\phi\to\bar{\phi}_R}\left|k_R(\theta_e)\frac{\partial\phi}{\partial\theta_e}\bigg|_{\theta_e\to\bar{\theta}_e}\right| + \frac{\partial T_a(\phi)}{\partial\phi}\bigg|_{\phi\to\bar{\phi}_L}\left|k_L(\theta_e)\frac{\partial\phi}{\partial\theta_e}\bigg|_{\theta_e\to\bar{\theta}_e}\right|$$

$$= w_R(\bar{\theta}_e)\frac{\partial T_a(\phi)}{\partial\phi}\bigg|_{\phi\to\bar{\phi}_R} + w_L(\bar{\theta}_e)\frac{\partial T_a(\phi)}{\partial\phi}\bigg|_{\phi\to\bar{\phi}_L}$$

$$\tag{9}$$

where $w_R(\bar{\theta}_e) = \left|k_R(\theta_e)\dfrac{\partial\phi}{\partial\theta_e}\bigg|_{\theta_e\to\bar{\theta}_e}\right|$, and $w_L(\bar{\theta}_e) = \left|k_L(\theta_e)\dfrac{\partial\phi}{\partial\theta_e}\bigg|_{\theta_e\to\bar{\theta}_e}\right|$ are positive constant values under a constant joint angle $\bar{\theta}_e$. Equation (9) suggests that the joint stiffness can be adjusted by setting $\bar{\phi}_R$ and $\bar{\phi}_L$ under holding Equation (6), and that the non-linearity of $T_a(\phi)$ is indispensable for the joint stiffness to be varied, because if $T_a(\phi)$ is linear with respect to ϕ, $\partial T_a(\phi)/\partial\phi$ takes a constant value, which means the stiffness S_e is also constant regardless of ϕ. $\bar{\phi}_R$ and $\bar{\phi}_L$ will be subject to gravitational effects as described above, which might narrow the adjustable range of stiffness. A weight compensator introduced in the upper arm, described in the following section, will alleviate this problem.

B. The case of r-ANLES

Stiffness regulation in the case of r-ANLES is simple. Figure 7 shows a configuration of a pair of r-ANLESes to control the rotation and stiffness of a rotary rod. The motors rotate guide shafts via gears, which twist torsion springs. One end of the springs is connected to the center disk that rotates together with the rotary rod. Therefore torques generated by the torsion springs are directly transmitted to the rotary rod. Torque loaded to the rotary rod is,

$$T_e = T_a(\phi_U) - T_a(\phi_L)$$

where, $T_a(\phi_U)$ and $T_a(\phi_L)$ are torques generated by upper-side and lower side r-ANLESes respectively. If we can neglect the gravitational effect, torsion angles ϕ_U and ϕ_L actually take the same value in an equilibrium state. Therefore, let us denote $\phi_U = \phi_L = \phi_e$

The stiffness around the rotary joint is then calculated by,

$$S_e = \left. \frac{\partial T_e}{\partial \theta} \right|_{\theta \to \theta_e} = \alpha \left. \frac{\partial T_e}{\partial \phi} \right|_{\phi \to \phi_e} = 2\alpha \left. \frac{\partial T_a(\phi)}{\partial \phi} \right|_{\phi \to \phi_e} \quad \text{[Nm/rad]}$$

where θ is the rotation angle of the rotary rod and α is a transmission ratio between θ and the torsion angle of r-ANLES ϕ (for the configuration shown in Figure 7, $\alpha = 1$).

3. TWO-DOF UPPER ARM

Figure 8a shows the structure of the upper limb. It has two DOFs: the flexion/extension of the elbow and the lateral/medial rotation of the upper arm. A pair of l-ANLESes controls the flexion/extension of the elbow. The lateral/medial rotation is controlled by a pair of r-ANLESes. Figure 8b shows the assembled view of the arm. The weight is about 6 kg. Table 1 shows the specifications of the motors and the spring used in the ANLESes.

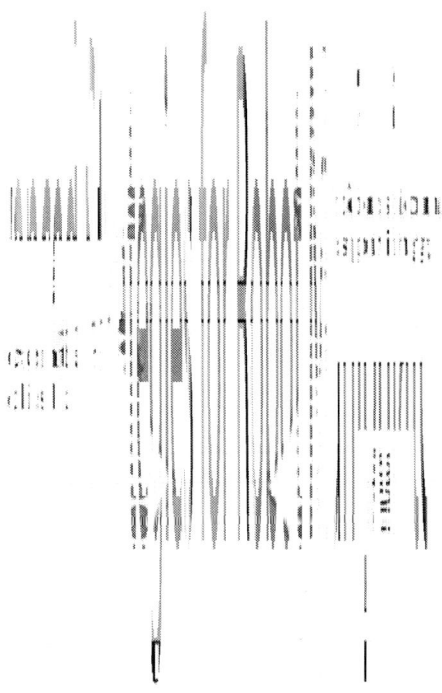

Figure 7. Rotary joint controlled by a pair of r-ANLES.

It is possible to use r-ANLESes for driving the flexion/extension by using bevel gears to change the rotary axes of the r-ANLESes. This requires a robust housing to sustain the axis rods and the bevel gears at both sides of the elbow joint, which might cause an increase in the weight and the moment of the arm's inertia.

Table 1. Specification of motors and springs used.

axis		Flexion/Extension	Lateral/Medial Rotation
Actuator (Motor)		Maxon RE25 × 2 Blushless DC 20 W Stall torque 250 mNm	Faulhaber 3257 × 2 Blushless DC 80 W Stall torque 540 mNm
Reduction gear ratio		132	134
Sensor (Rotary Encoder)		Maxon HEDS 5540 × 2	Faulhaber IE2-16 × 2
spring	modulus of longitudinal elasticity	1.86×10^{5} [N/mm^{2}]	
	Wire diameter	1.9 mm	4.0 mm
	Coil diameter	32.7 mm	36 mm
	Number of coiling	24	15
	pitch	2.0 mm	4.2 mm

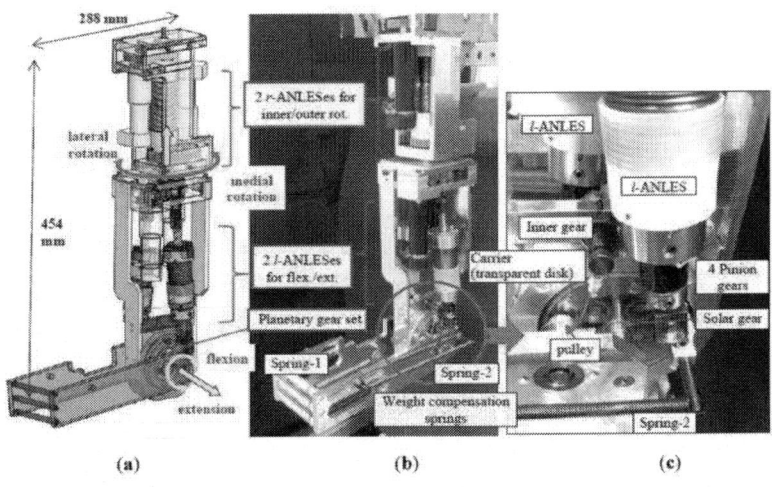

(a)   (b)   (c)

Figure 8. (**a**). Structure of the two- degree-of-freedom (DOF) upper arm; (**b**). Assembled view of the two-DOF upper arm; (**c**). Assembled view of the elbow joint part.

3.1. Elbow Joint

The elbow joint mainly consists of a planetary gear set (PGS) (see Figure 8c). The inner gear of the PGS is connected to the upper arm and the solar gear is connected to the forearm. Two carriers are equipped at both sides of the

PGS, which support the four rotary axes of the pinion gears. Each carrier is connected to the l-ANLES, on which connecting points are located so that two l-ANLESes work counteractively, as shown in Figure 8a,b. The rotation of the motor of l-ANLES in one side in the opposite direction to that of the motor in other side brings about prismatic motions of two ball screw rods in mutually opposite directions, providing a rotation of the carrier (transparent disks shown in Figure 8c) of at most 52 deg. However, it is transmitted to the solar gear's rotation (and, therefore, the forearm's rotation) about three times larger due to the speed-up ratio of the PGS (teeth number of the inner gear/teeth number of the solar gear), resulting in about a 160-degree rotation of the elbow joint, which is almost equal to the rotation range of a human elbow (approximately 140° [35]). The rotation of the l-ANLES motors by the same angle in the same direction provides the same amount of traction force of the ball screw rods, which induces the same amount of torque in the opposite direction of the elbow joint because of the antagonistic configuration of the two l-ANLESes. Therefore, they counterbalance without the elbow joint's rotation. However, concomitantly, the torsion spring of each l-ANLES wraps around the guide shaft by the motor rotation, which augments the stiffness of the elbow joint, as verified in the experiments described below.

3.2. Lateral/Medial Rotation of the Upper Arm

A pair of r-ANLESes controls the lateral/medial rotation of the upper arm (see Figure 3, Figure 8a,b). The rotations of the r-ANLES motors in mutually opposite directions bring about the lateral/medial rotation of the arm, of which the rotation angle is not limited unless the motion of the forearm or the upper arm is not hampered by obstacles. The rotations in the same direction twist the two torsion springs in mutually opposite directions, enhancing the stiffness around the lateral/medial axis, as verified by the experiments described below.

3.3 Weight Compensator

The weight of the forearm is compensated by the weight compensation springs shown in Figure 8a and Figure 8b. The weight compensation is indispensable because if it is absent, l-ANLESes have to sustain all of the forearm weight. This requires some pre-torsion of the springs of the l-ANLES, the amount of which depends on the elbow angle. Therefore, a lack of the weight compensator will deteriorate the controllability of the joint stiffness by l-ANLES.

Figure 9a shows the spring alignment. Spring-1 is connected between the upper arm and the forearm via the pulley as shown in Figure 8b,c. Spring-2 is connected between the forearm and the inner gear of the PGS as also shown in Figure 8b,c. A successful aspect of this weight compensation mechanism resides in the fact that the inner gear rotates in the opposite direction to the solar gear's rotation (therefore, the forearm's rotation). Hence, the connecting point p at the inner gear of Spring-2 transfers, as shown in Figure 9a, so that it passes to the axis of the elbow joint, which changes the sign of the loaded torque due to Spring-2.

Figure 9b shows the experimental apparatus to the load torque that is necessary for the elbow joint to take a specified angle. Three weights pull the distal end of the forearm via wires and pulleys to give an appropriate torque about the elbow joint in any joint angle. Procedures of the experiments are as follows.

1. The weights w_1, w_2 and w_3 are adjusted so that the elbow joint takes a specified joint angle θ_e that is measured by the optical encoder attached to the rotary axis of the elbow joint.

2. Then, the torque loaded around the elbow joint is measured by

$$\tau_e = r_x(f_{1y} + f_{2y} + f_{3y}) - r_y(f_{1x} + f_{2x} + f_{3x})$$

where, $\mathbf{r} = [r_x\ r_y]^T$ is a vector from the elbow joint to the distal end of the forearm, and $\mathbf{f}_i = [f_{ix}\ f_{iy}]^T$, ($i = 1,2,3$) is force vectors generated by the weights.

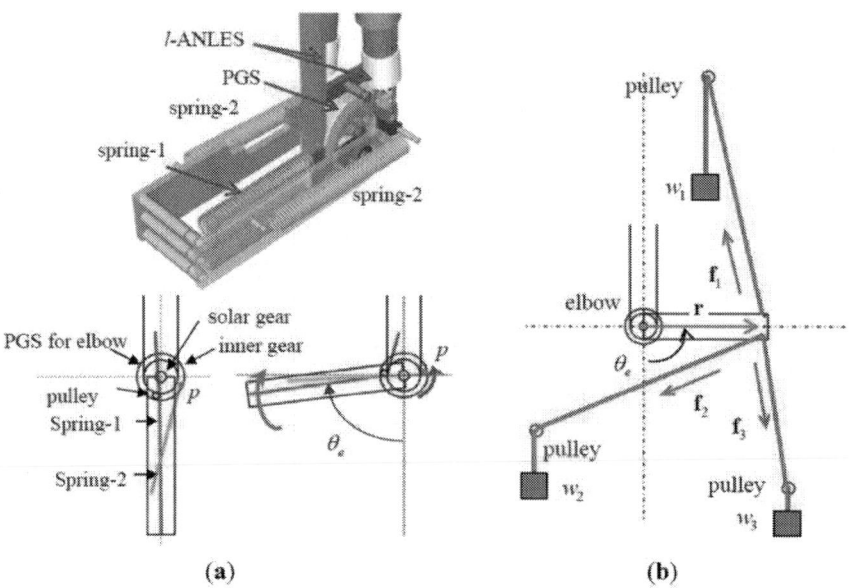

Figure 9. (a) Placement of the weight compensation springs. (b) Experimental apparatus for the weight compensator.

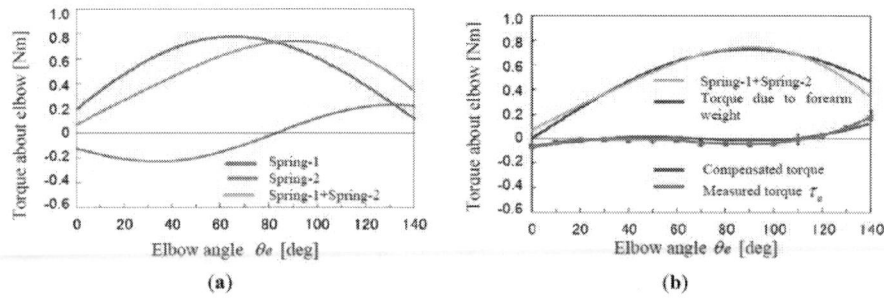

Figure 10. Theoretical and Experimental result of weight compensation of the forearm. (a) Torques at the elbow joint due to springs; (b) Torques at the elbow joint due to springs, forearm weight and their compensating weights.

As shown in Figure 10a, the torque exerted by Spring-2 changes from a negative value to a positive one as the elbow flexes, since Spring-2 crosses over the elbow axis due to the rotation of the inner gear. Hence, the resultant torque of Spring-1 and Spring-2 effectively compensates the torque due to the forearm weight, as shown with the line labeled "Compensated torque" in Figure 10b. The compensated torque takes a nearly zero value throughout the whole range of the elbow angle, which is verified by measuring the torque of the elbow joint as shown in Figure 10b. Figure 10b also shows error bars of the data variation of five trials of identical experiments, but they are less visible because of such good repeatability of the experiments.

Figure 10b also shows that the torque due to the weight of the forearm is not entirely compensated, especially in the ranges of the elbow joint angle less than 20° and more than 120°. These residual torques are loaded on a l-ANLES and required some pre-torsion of the torsion spring of the l-ANLES to hold the specified elbow joint angle. However, it does not harm the performance of the arm as a whole because the value of the non-compensated torque is so small, and because it would be uncommon for the arm to do dexterous jobs requiring stiffness regulation in such ranges of the elbow joint angles.

The lateral/medial joint has not yet been equipped with a weight compensator, which will be required when the arm is connected to a shoulder joint of which the rotation will engender a torque due to gravitational force around the lateral/medial joint of the arm. This will also require a change in the weight compensation around the elbow joint according to the shoulder joint angles.

3.4. Stiffness Control of the Flexion/Extension Axis and the Lateral/Medial Rotation Axis

The stiffness of the flexion/extension axis of the elbow joint and the lateral/medial rotation axis of the upper arm can be controlled in an antagonistic action of a pair of l-ANLESes and r-ANLESes, respectively. The same amount of twisting of the torsion springs in both ANLESes changes the stiffness of the joint. Figure 11 and Figure 12 show the results of the experimental measurement of the stiffness accompanied by a theoretical curve. "Torsion angle of ANLES" in the horizontal axis refers to the same amount of rotation angle of the motors of a pair of ANLESes rotating the guide-shafts.

The experimental procedures are as follows:

1. The non-linear elasticity of the ANLES obeys the one shown in Figure 5.

2. Two motors of the l-ANLESes are rotated so that the elbow angle reaches 90° and two motors of the r-ANLESes are rotated so that the lateral/medial rotation angle reaches 0 deg. The motors of the l-ANLESes and r-ANLESes are controlled by a normal PID controller.

3. In the above state, the torsion angles of both l-ANLESes and r-ANLESes are almost zero because the pre-torsion of ANLESes to sustain the forearm weight is not needed due to the weight compensator.

4. The motors of the l-ANLESes or the r-ANLESes are rotated by the same angle to give the torsion springs of the l-ANLESes or the r-ANLESes the same amount of torsion angle in order to increase the stiffness of the elbow or the lateral/medial joint.

5. A weight W was loaded at the forearm tip downward in the case of the elbow joint, or horizontally in the case of the lateral/medial rotation axis. The weight was adjusted to attain a 5-degree rotation of the joint that is measured by the optical encoder and displayed on the computer screen in each setting of the torsion angle of ANLESes. Then, the stiffness is obtained as: (Nm/rad) with the forearm length l.

The load torque is hardly changed by the 5-degree rotation of the joints (only 0.4% reduction). This reduction is completely decreased when a smaller rotation angle is given. However, the load torque becomes closer to the friction torque of the joints if a smaller rotation angle is given, bringing about a wide variation of stiffness data.

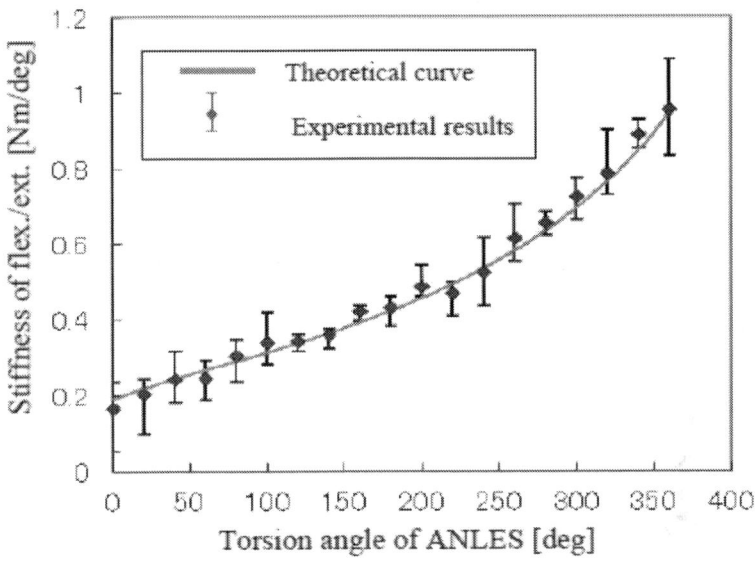

Figure 11. Stiffness about flexion/extension of the elbow joint.

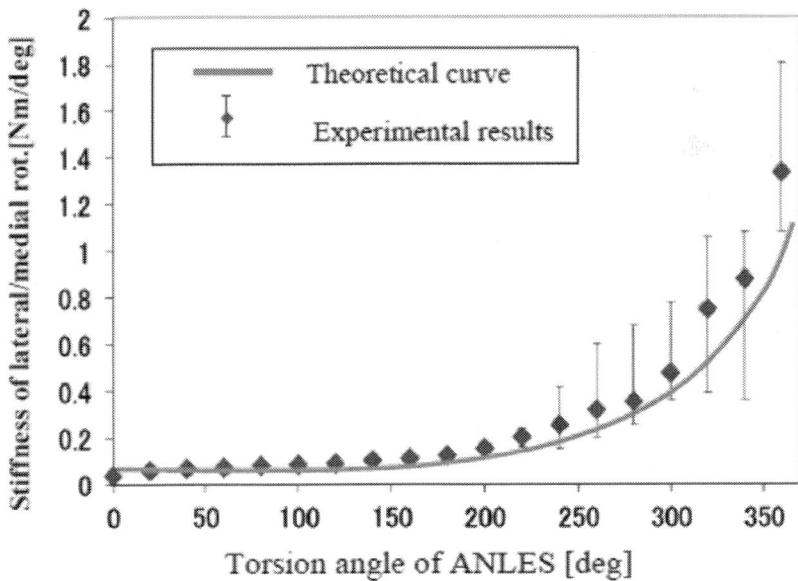

Figure 12. Stiffness about the lateral/medial rotation of the upper-arm.

We achieved the same experimental results five times according to the procedure described above. The error bars in Figure 11 and Figure 12 show their data variation, as well as the measured stiffness, which shows much variation as the torsion angle of ANLESes increases. This is due to friction yielding between the torsion spring and the guide-shaft of ANLESes. When the spring is twisted and coiled on the guide-shaft, the spring wire has to be moved slightly on the guide-shaft because the pitch narrows.

The result shown in Figure 11 and Figure 12 elucidates the validity of our theory since the experimental data coincided well with the theoretical curve. The cause of the experimental data being a little higher than the theoretical curve is due to the friction that emerged during torque transmission with gears or bearings.

4. CONCLUSIONS

In this paper, a two-DOF upper arm with a musculo-skeletal structure is introduced, in which the ANLES works as a voluntary muscle. Two types of ANLESes are used for controlling the joint angle and joint stiffness in the flexion/extension of the elbow joint and lateral/medial flexion of the upper arm: the linear type ANLES (l-ANLES) and the rotary type ANLES (r-ANLES), respectively. Combined with our previous study in which two types of ANLES were used in a wrist joint [31,32,33], the present study clarifies that ANLES can be used as actuators in a wide range of robots driven on the basis of antagonistic actuations.

Advantages of using ANLES are as follows:

(1) The magnitude and the adjustable range of joint stiffness can be precisely designed by designing the shape of the guide-shaft.

(2) No particular kind of non-linear elastic elements are necessary to construct the ANLES.

We are now constructing a two-DOF shoulder joint that is also driven by ANLESes and plan to combine the upper arm to constitute a four-DOF arm. Subsequently, we plan to combine the four-DOF arm with the three-DOF wrist joint that has been completed to construction [34], which will provide a seven-DOF manipulator controlled in an antagonistic manner similar to the human upper extremity.

ACKNOWLEDGMENTS

The authors thank the Japan Society for the Promotion of Science, via Grants-in-Aid for Scientific Research: No. 18560258 "Stiffness control of an antagonistically driven joint using the actuator with non-linear elasticity" for their support in the pursuance of this work.

REFERENCES

1. Shadmehe, R.; Arbib, M.A. A mathematical analysis of the force-stiffness characteristics of muscles in control of a single joint system. Biol. Cybern. **1992**, 66, 463–477.

2. Matthews, P.B.C. The dependence of tension upon extension in the stretch reflex of the soleus muscle of the decerebrated cat. J. Physiol. **1959**, 147, 521–546.

3. Fel'dman, A.G. Functional tuning of the neurons system with control of movement or maintenance of a steady posture. Biofizika **1966**, 11, 498–508.

4. Hoffer, J.A.; Andrearsen, S. Regulation of soleus muscle stiffness in premammiliary cats. J. Neurophysiol. **1981**, 45, 267–285.

5. Dolan, J.M.; Friedman, M.B.; Nagurka, M.L. Dynamic and Loaded Impedance of Human Arm Posture. IEEE Trans. Syst. Man Cybern. **1993**, 23, 698–709.

6. Mussa-Ivaldi, N.H.; Bizzi, E. Neural, Mechanical, and Geometrical Factors Subserving Arm Posture in Humans. J. Neurosci. **1985**, 5, 2732–2743.

7. Jocobsen, S.C.; Wood, J.E.; Knutti, D.F.; Biggers, K.B. The UTAH/M.I.T. Dexterous Hand: Work in Progress. Int. J. Rob. Res. **1984**, 3, 21–51.

8. Jacobsen, S.C.; Ko, H.; Inversen, E.K.; Davis, C.C. Antagonistic Control of a Tendon Driven Manipulator. In Proceedings of the 1989 IEEE International Conference on Robotics and Automation, Scottsdale, AZ, USA, 14–19 May 1989.

9. Laurin-Kovitz, K.F.; Colgate, J.E.; Carnes, S.D.R. Design of Components for Programmable Passive Impedance. In Proceedings of the 1991 IEEE International Conference on Robotics and Automation, Sacramento, CA, USA, 9–11 April 1991.

10. Yi, B.J.; Freeman, R.A. Geometric Characteristics of Antagonistic Stiffness In Redundantly Actuated Mechanisms. In Proceedings of the 1993 IEEE International Conference on Robotics and Automation, Atlanta, GA, USA, 2–6 May 1993.

11. Kobayasi, H.; Hyoudou, K.; Ogane, D. On Tendon-Driven Robotics Mechanisms with Redundant Tendons. Int. J. Rob. Res. **1998**, 17, 561–571.

12. Lee, Y.T.; Choi, H.R.; Chung, W.K.; Youm, Y. Stiffness Control of a Coupled Tendon-Driven Robot Hand. IEEE Contr. Syst. Mag. **1994**, 14, 10–19.

13. Yi, B.J.; Freeman, R.A. Synthesis of Actively Adjustable Springs by Antagonistic Redundant Actuation. J. Dyn. Syst. Meas. Contr. **1992**, 114, 454–461.

14. Chen, S.F.; Kao, I. Conservative Congruence Transformation for Joint and Cartesian Stiffness Matrices of Robotic Hands and Fingers. Int. J. Rob. Res. **2000**, 19, 835–847.

15. Tondu, B.; Lopez, P.M. Modeling and Control of McKibben Artificial Muscle Robot Actuators. IEEE Contr. Syst. Mag. **2000**, 20, 15–28.

16. Tondu, B.; Ippolito, S.; Guiochet, J.; Daidie, A. A Seven-degrees-of-Freedom Robot-arm Driven by Pneumatic Artificial Muscles for Humanoid Robots. Int. J. Robot. Res. **2005**, 24, 257–274.

17. Verrelst, B.; Van Ham, R.; Vanderborght, B.; Daerden, F.; Lefeber, D.; Vermeulen, J. The Pneumatic Biped "Lucy" Actuated with Pleated Pneumatic Artificial Muscles. Autonom. Rob. **2005**, 18, 201–213.

18. Koganezawa, K.; Watanabe, Y.; Shimizu, N. Antagonistic Muscle-Like Actuator and Its Application to Multi-DOF Forearm Prosthesis. Adv. Rob. **1999**, 12, 771–789.

19. Migliore, S.A.; Brown, E.A.; DeWeerth, S.P. Biologically Inspired Joint Stiffness Control. In Proceedings of the 2005 IEEE International Conference on Robotics and Automation, Barcelona, Spain, 18–22 April 2005; pp. 4508–4517.

20. Tonietii, G.; Schiavi, R.; Bicchi, A. Design and Control of a Variable Stiffness Actuator for Safe and Fast Physical Human/Robot Interaction. In Proceedings of the 2005 IEEE International Conference on Robotics and Automation, Barcelona, Spain, 18–22 April 2005; pp. 526–531.

21. Schiavi, R.; Grioli, G.; Sen, S.; Bicchi, A. VSA-II: A Novel Prototype of Variable Stiffness Actuator for Safe and Performing Robots Interacting with Humans. In Proceedings of the 2008 IEEE International Conference on Robotics and Automation, Pasadena, CA, USA, 19–23 May 2008; pp. 2171–2176.

22. Wolf, S.; Hirzinger, G. A New Variable Stiffness Design: Matching Requirements of the Next Robot Generation. In Proceedings of the IEEE International Conference on Robotics and Automation, Pasadena, CA, USA, 19–23 May, 2008; pp. 1741–1746.

23. Haddadin, S.; Albu-Schaffer, A.; Hirzinger, G. The Role of the Robot Mass and Velocity in Physical Human-Robot Interaction-Part 2: Constrained Blunt Impacts. In Proceedings of the IEEE International Conference on Robotics and Automation (ICRA2008), Pasadena, CA, USA, 19–23 May 2008; p. 1339.

24. Ham, R.; Sugar, T.G.; Vanderborght, B.; Hollander, K.W.; Lefeber, D. Compliant Actuator Design. IEEE Rob. Autom. Mag. **2009**, 16, 81–94.

25. Ham, V.; Vanderborght, B.; Van Damme, M.; Verrelst, B.; Lefeber, D. MACCEPA, the Mechanically Adjustable Compliance and Controllable Equilibrium Position Actuator: Design and Implementation in a Biped Robot. Rob. Autonom. Syst. **2007**, 55, 761–768.

26. Schiavi, R.; Grioli, G.; Sen, S.; Bicchi, A. VSA-II: A Novel Prototype of Variable Stiffness Actuator for Safe and Performing Robots Interacting with Humans. In Proceedings of the IEEE International Conference on Robotics and Automation, Pasadena, CA, USA, 19–23 May 2008; pp. 2171–2176.

27. Wolf, S.; Hirzinger, G. A New Variable Stiffness Design: Matching Requirements of the Next Robot Generation. Proceedings of the 2008 IEEE International Conference on Robotics and Automation 1741–1746.

28. Byeong-Sang, K.; Song, J. Hybrid Dual Actuator Unit: A Design of a Variable Stiffness Actuator Based on an Adjustable Moment Arm Mechanism. Proceedings of the 2010 IEEE International Conference on Robotics and Automation 1655–1660.

29. Vanderborght, B.; Albu-Schaeffer, A.; Bicchi, A.; Caldwell, D.; Tsagarakis, N.; Van Damme, M.; Lefeber, D.; Van Ham, R.; Burdet, E.; Carloni, R.; et al. Variable Impedance Actuators: A Review. Rob. Autonom. Syst. **2013**, 61, 1601–1614.

30. Morita, T.; Sugano, S. Design and Development of a New Robot Joint Using a Mechanical Impedance Adjuster. In Proceedings of the IEEE International Conference on Robotics and Automation, Nagoya, Japan, 21–27 May 1995; pp. 2469–2475.

31. Van Ham, R.; Van Damme, M.; Verrelst, B.; Vanderbought, B.; Lefeber, D. MACCEPA, The Mechanically Adjustable Compliance and Controllable Equilibrium Position Actuator: A 3DOF Joint with Two Independent Compliances. Int. Appl. Mech. **2007**, 43, 467–474.

32. Koganezawa, K. Mechanical Stiffness Control for Antagonistically Driven Joints. In Proceedings of the IEEE/RSJ International Conference

on Intelligent Robots & Systems (IROS 2005), Edmonton, Canada, 2–6 August 2005; pp. 2512–2519.

33. Koganezawa, K.; Yamashita, H. Stiffness Control of Multi-DOF Joint. In Proceedings of the 2009 IEEE/RSJ International Conference on Intelligent Robots and Systems (IROS 2009), St.Louis, USA, 11–15 October 2009; pp. 363–370.

34. Koganezawa, K.; Takami, G.; Watanabe, M. Antagonistic Control of Multi-DOF Joint. In Proceedings of the 2012 IEEE/RSJ International Conference on Intelligent Robots and Systems (IROS 2012), Vilamoura, Portugal, 7–12, October 2012; pp. 2895–2900.

35. Watkins, J. Structure and Function of the Musculoskeletal System; Human Kinetics: Champaign, IL, USA, 1999; p. 186.

CHAPTER 9

Affine Transform to Reform Pixel Coordinates of EOG Signals for Controlling Robot Manipulators Using Gaze Motions

Muhammad Ilhamdi Rusydi [1,2], Minoru Sasaki [1,*] and Satoshi Ito [1]

[1] Department of Mechanical Engineering, Gifu University, 1-1 Yanagido, Gifu City, 501-1193, Japan
[2] Department of Electrical Engineering, Andalas University, Limau Manis, Padang City, 25163, Indonesia

ABSTRACT

Biosignals will play an important role in building communication between machines and humans. One of the types of biosignals that is widely used in neuroscience are electrooculography (EOG) signals. An EOG has a linear relationship with eye movement displacement. Experiments were performed to construct a gaze motion tracking method indicated by robot manipulator movements. Three operators looked at 24 target points displayed on a monitor that was 40 cm in front of them. Two channels (Ch1 and Ch2) produced EOG signals for every single eye movement. These signals were converted to pixel units by using the linear relationship between EOG signals and gaze motion distances. The conversion outcomes were actual pixel locations. An affine transform method is proposed to determine the shift of actual pixels to target pixels. This method consisted of sequences of five geometry processes, which are translation-1, rotation, translation-2, shear and dilatation. The accuracy was approximately $0.86° \pm 0.67°$ in the horizontal direction and $0.54° \pm 0.34°$ in the vertical. This system successfully tracked the gaze motions not only in direction, but also in distance. Using this system, three operators could operate a robot manipulator to point at some targets. This result shows that the method is reliable in building communication between humans and machines using EOGs.

KEYWORDS

EOG; gaze motions; affine transform; linear relationship; actual pixels; target pixels; robot manipulator

1. INTRODUCTION

The electrooculography (EOG) signal is a bio-signal that measures eye activities. Those activities generate a potential difference between the cornea and the retina. Although many methods can learn the phenomena of eye movement, the potential difference [1] is the most broadly used by neuroscientists to investigate eye movements [2]. The EOG signal has a linear characteristic with regard to the gaze motion distance [3]. This linear condition of the EOG happens at approximately $\pm45°$ in the horizontal direction and $\pm30°$ in the vertical [4].

Some methods have been developed to determine the relationship between the EOG and gaze motions. Linear predictive coding cepstrum (LPPC) was used by [5] as the feature for eye movement pattern matching. In various experiments, this research also used dynamic time warping to compensate the EOG period. A spectral entropy algorithm was implemented to detect the endpoints of the EOG to improve the accuracy of the recognition system among noisy signal conditions. The purpose of the experiments in this study was to detect seven eye activities when looking up, down, left, right, blinking two times, blinking three times and blinking four times.

Barea et al. [6] proposed a system to identify horizontal gaze motions based on a neural network. They obtained ±10° with a 2° error accuracy for horizontal gaze motion. A neural network for EOG signal recognition was also implemented to help physicians make a diagnosis to distinguish normal and subnormal eye conditions [7]. A neural network for classifying the EOG in six conditions was also performed by [8]. The conditions that were distinguished were straight, up, down, right, left and blink. Combinations between three EOG features, which were the wavelet detail coefficient, power spectral density and auto regressive coefficient, were also performed to compare their accuracy.

Fuzzy logic was implemented for detecting eye movement by using the EOG in [9]. The horizontal EOG was divided into four groups: right, left, hard right and hard left. The vertical EOG was grouped into up and down only.

A low-cost computer interface based on the EOG was proposed by [10]. This interface used the polarity of signals to differentiate the gaze motion direction on each channel. The peak amplitude and the slope of the signals were used to categorize the blink, eye movement and noise. This method successfully recognized right, left, up and down gaze motion and blinks.

In [11], the integral value of the EOG was introduced as a feature to detect eye movement. Two interface models were controlled by three operators. The first had eight options and the other had 12 options for gaze motion. This method showed better accuracy for both models as compared to the maximum EOG value method.

Several researchers have already invented some applications. Pinhero et al. [12] reviewed EOG functions to help disabled people communicate with other people through a machine controlled by EOG. EOG signals were converted into alphanumeric/symbol/number and cursor control signals and also used to generate Morse code. EOG was also used to operate a TV and play games [13].

This paper proposes an affine transform method to build a gaze motion tracking system. A homogeneous affine transform was constructed by sequences of five geometry processes, which are translation-1, rotation, translation-2, shear and dilation. This method was designed to detect the direction and the distance of gaze motions. Three operators tested the system by gazing at 24 target points. Finally, a robot manipulator was used as the indicator of gaze motions to some target points.

2. EXPERIMENTAL ENVIRONMENT

2.1. EOG Signal

This research focused on building an eye tracking system of direction and distance of gaze motions using EOG signals. An instrument produced by the NF Corporation (Yokohama, Japan) was used to measure the EOG. This sensor has four electrodes, a processor box and a head box as the system amplifier, as shown in Figure 1a. A 60 Hz low-pass filter, as shown by Equation (1), was used inside the processor box, since standard electric noise occurred even when the electrodes were not attached on the skin. Elefix paste (Nihon Koden, city, country), a highly conductive and low impedance gel as shown by Figure 1b, stuck the electrodes to the dry skin, so the electrodes were in stable positions:

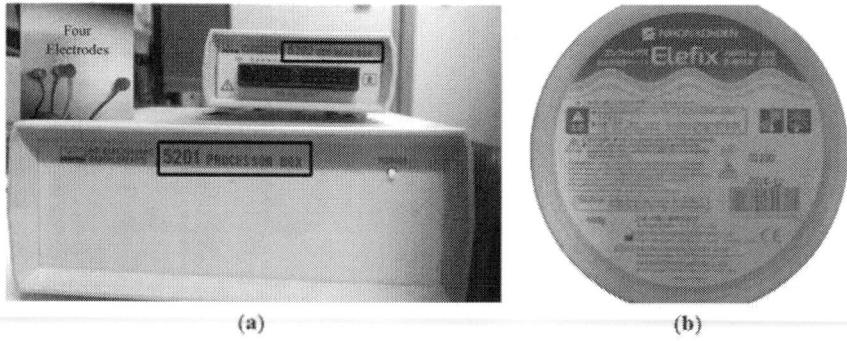

(a)   (b)

Figure 1. (a) NF instrument for EOG sensors with a processor box, a head box and four electrodes. (b) Elefix paste produced by Nihon Koden.

$$H(s) = \frac{4\pi^2 f^2}{s^2 + 4\pi f \cos\left(\dfrac{\pi}{4}\right) + 4\pi^2 f^2}$$

$f = 60\,Hz\ cutoff\ frequency$

(1)

The electrodes consisted of a ground channel, a reference channel, channel 1 (Ch1) and channel 2 (Ch2). Ch1 detected the signal for vertical gaze motions and Ch2 for horizontal gaze motions. EOG is very sensitive to electrode position and many different methods to record the EOG signal are used in different laboratories [14]. In [15], some electrode position possibilities, as shown by Figure 2, were checked based on average and standard deviation of the integral EOG for some gaze motion distances. The result showed that electrode position number 2 for horizontal gaze motions (Ch2) and number 11 (Ch1) for vertical gaze motions had a stable EOG signal and linear relationship with gaze distance.

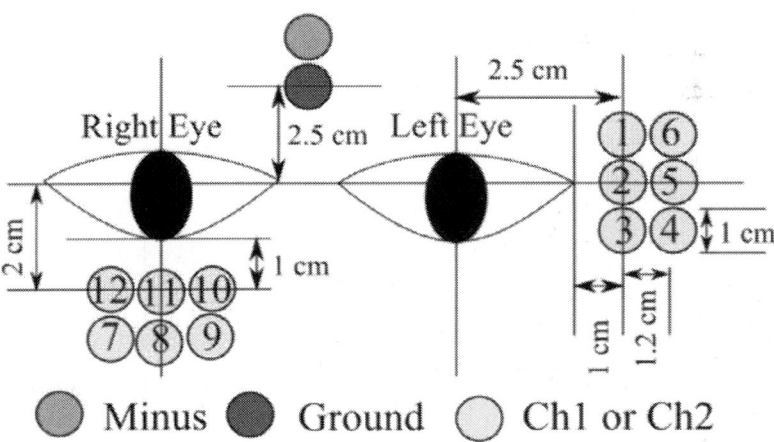

Figure 2. Possibilities of electrode positions.

Three features of the EOG signal were used in this study. First were the threshold values. These were types: the positive threshold (Th+) and the negative threshold (Th−). The signals between these thresholds indicated condition of eyes (at rest/no movement). Second was the polarity of the signal from Ch1 and Ch2. The polarity was defined from which threshold (Th+ or Th−) was passed first by the EOG. Positive signal (+) was the condition for EOG signal if it first passed the positive threshold and negative polarity was the condition for EOG signal if it first passed the negative threshold.

If the polarity of the signal of Ch1 was negative (−) and the polarity of the signal of Ch2 was negative (−), they were grouped into Area 1. In Area 2, the polarity of the signal of Ch1 was positive (+) and the polarity of the signal of Ch2 was negative (−). In Area 3, both of the polarities were positive (+). In Area 4, the polarity of the signal of Ch1 was positive (+) and the polarity of the signal of Ch2 was negative (−). Figure 2 illustrates these conditions and Table 1 lists them.

Table 1. Combinations of EOG Signals.

	Area 1	Area 2	Area 3	Area 4
Ch1	Negative (−)	Positive (+)	Positive (+)	Negative (−)
Ch2	Negative (−)	Negative (−)	Positive (+)	Positive (+)

The last EOG feature in this experiment was its integral, which had a linear relationship with gaze distance. In Figure 3, the green shadow for the EOG signal from Ch1 in Area 2 shows an example of integral of EOG. The integral was calculated from the first time the signal passed the zero value until it constructed a full wave, as shown by Equation (2). The integral of EOG was normalized to scale the data into 0 to 1. The linear relationship between normalized of integral EOG and eye distance is shown by Equation (3):

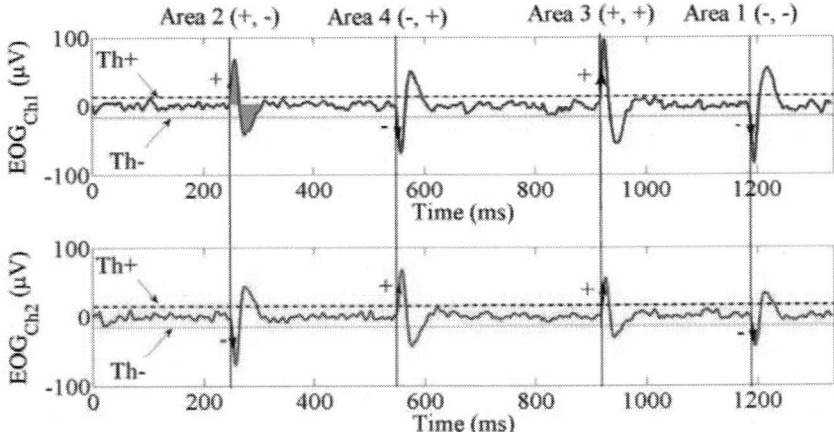

Figure 3. EOG signals for Areas 1, 2, 3 and 4.

$$int\_EOG_{Chi} = \left| \int_{\Omega_+} EOG_{Chi}(t)\,dt \right| + \left| \int_{\Omega_-} EOG_{Chi}(t)\,dt \right|$$

$$\Omega_+ = \{t : EOG_{Chi}(t) > th+\}$$

$$\Omega_- = \{t : EOG_{Chi}(t) < th-\}$$

$$i = 1,2 \tag{2}$$

$$Dis\tan ce = A*norm\_int\_EOG_{Chi} + B$$
$$i = 1\ for\ up\ or\ down\ (v)\ and\ 2\ for\ right\ or\ left\ gaze\ motion\ (u)$$
$$Distance = v\ pixel\ for\ up\ or\ down\ and\ u\ pixel\ for\ right\ or\ left\ gaze\ motion$$
$$norm\_int\_EOG_{Chi} = normalization\ of\ integral\ EOG \tag{3}$$

In the previous research [16], some horizontal and vertical gaze distances were measured to determine variables A and B. Some horizontal and vertical targets were put on a monitor which was 40 cm in front of operators. The size of the monitor was 34 cm × 27 cm (width × height). The distance between two targets in horizontal was 255 pixels and the distance between two targets in the vertical was 180 pixels. The average values of A and B for the four basic directions (up, down, right and left) of gaze motion were 900 and −300 for up, −850 and 200 for down, 1,100 and −138 for right and −1,100 and 130 for left gaze motion. The vertical gaze motions (up and down) were associated with movement in v pixel and the horizontal gaze motions were organized into the u pixel. Figure 4a shows the signal from Ch1 for four gaze distances (180, 360, 540, and 720 pixels) in vertical gaze motions (down and up) and Figure 4b illustrates the signal from Ch2 for four gaze distances (255, 510, 765, and 1,020 pixels) in horizontal gaze motions (right and left).

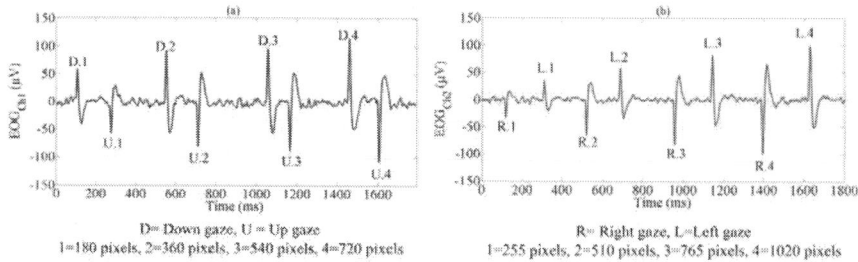

Figure 4. Four kind of gaze motion distances (**a**) EOG signal from Ch1 in vertical direction, and (**b**) EOG signal from Ch2 in horizontal direction.

2.2. Integrated System

Figure 5 illustrates the hardware setup of this experiment. There were two target plates, one is for gaze targets and another is for the robot targets. An EOG program worked with the input was EOG signals and the output was the angle degrees for the robot manipulator.

There were three system coordinates in this research, target coordinate (u,v) in pixel, target coordinate (x,y) in cm and robot coordinate (x_R, y_R) in cm. Because of the offside positions between (x,y) and (x_R, y_R), the relationship between (x,y) and (P_x, P_y) was calculated by Equation (4):

$$(Px, Py) = (x - 23cm, \ y + 27cm)$$

(4)

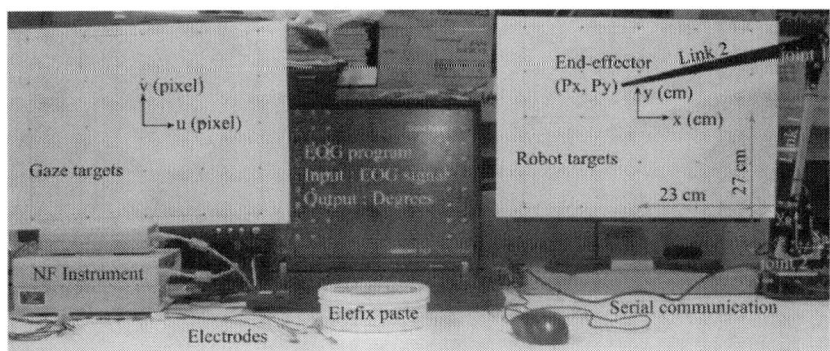

Figure 5. Experiment setup of robot control system using EOG.

A planar robot manipulator was used as indicator of eye movements. It had two joints with length of both joints was same, 30 cm. Angle of joint 1 was represented by α and angle of joint 2 was named by β. The area of α was from 0° to 180° and the area of β was from 0° to 140°. The relationship between end-effector position (P_x, P_y) and the joint angles are given by Equations (A1)–(A7) in Appendix A. This robot was connected using serial communication between an Arduino microcontroller and a computer. This computer received EOG data from the processor box.

3. METHODOLOGY

Three human operators attempted to track their gaze motion by using EOG in this research. Their heads were fixed 40 cm in front of a monitor, where the target points appeared. A sequence of experiments was conducted to find the gaze motion pixel coordinates or actual pixel positions. A pixel coordinate was evaluated by comparing it to the target pixel position. The geometry process was used to improve the performance of the gaze motion tracking system.

3.1. Training Targets

In total, 24 training targets used in this research were symmetrically spread across the monitor, as shown in Figure 6. The size of the monitor was 1,020 pixels × 720 pixels or 34 cm × 27 cm (horizontal × vertical). Every pixel in the horizontal and vertical could be converted to cm unit by Equation (5). The monitor was 40 cm in front of operators. Using trigonometry rules, the 255 pixels in horizontal directions were equal to 12° and 180 pixels in vertical directions was approximately 9°. The training targets were named based on their location. Table 2 lists the target names and their pixel positions. The areas of the targets were based on the EOG signal type:

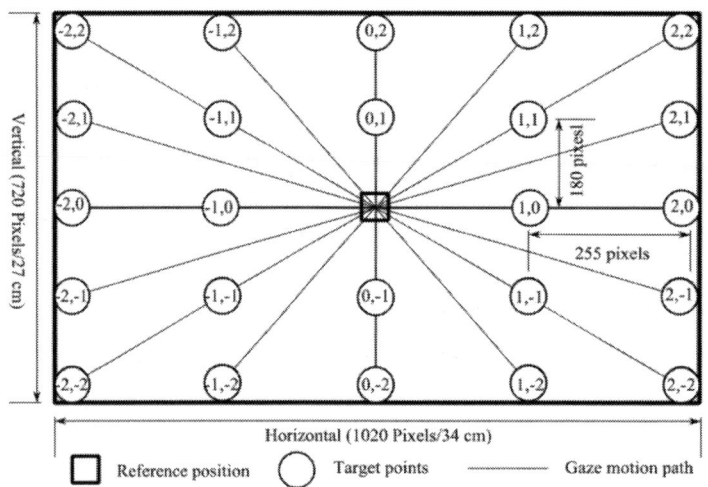

Figure 6. Training targets and eye reference positions.

Table 2. Pixel locations of each training target.

Area 1			Area 2			Area 3			Area 4		
Target	Pixel		Target	Pixel		Target	Pixel		Target	Pixel	
	x	y		x	y		x	y		x	y
0,1	0	180	1,0	255	0	0, −1	0	−180	−1,0	−255	0
0,2	0	360	1,−1	255	−180	0, −2	0	−360	−1,1	−255	180
1,1	255	180	1,−2	255	−360	−1, −1	−255	−180	−1,2	−255	360
1,2	255	360	2,0	255	0	−1, −2	−255	−360	−2,0	−255	0
2,1	510	180	2, −1	510	−180	−2, −1	−510	−180	−2,1	−510	180
2,2	510	360	2, −2	510	−360	−2, −2	−510	−360	−2,2	−510	360

$$(x,y) = \left(\frac{u*34}{1020} cm, \frac{v*27}{720} cm \right)$$

(5)

The three operators performed gaze motions from the reference position to the target points. A reference point was the (0,0). Each movement was done five times for a total of 120 gaze motions.

3.2. Reference Line

Figure 7 shows the patterns of the actual pixel positions in the four areas. The green squares are the target pixel positions. The red circles are the actual pixels from Ch1 and Ch2. Each area had six target positions and six actual positions.

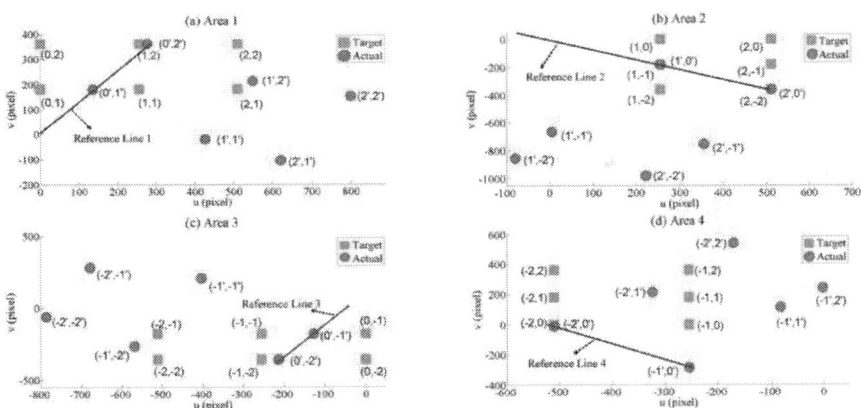

Figure 7. Actual and target pixels from the EOGs.

Each area had a reference line. A reference line was needed since the rotation geometry process is the projection of the actual pixels to this line. In addition, these lines were also necessary to perform the shear geometry process. At a glance, the actual pixels could be grouped into two patterns. Area 1 had reference line 1, which was a linear line drawn between the actual pixels at (0',1') and (0',2'). Area 3 had the same pattern as area 1, but reference line 2 was drawn between the actual pixels at (0',−1) and (0,−2). Area 2 and 4 drew their reference lines from actual pixels that occurred from the horizontal gaze motion. In this case, reference line 2 was drawn between the actual pixels at $(1',0')$ and $(2',0')$, whereas the actual pixels at $(-1',0')$ and $(-2',0')$ composed reference line 4. Since the reference lines are linear, they are simply written by Equation (6):

$$y = a_i x + b_i, i = area\,(1; 2; 3; or\,4).$$

(6)

3.3. Homogeneous Matrix of the Affine Transform

A homogeneous matrix of the affine transform, based on the basic affine transform in Appendix B, was proposed to match the actual pixels with the target pixels in this research to compensate the translation processes, which are not linear with other translation processes [17]. Each of the four areas had a homogeneous matrix. In this study, the matrices were built from the sequences of the five geometry processes. Figure 8 illustrates how these processes worked.

Figure 8a illustrates the positions of the four actual pixels before the transformation in comparison to the target points. The translation process in Figure 8b was fundamental since the actual pixel was rotated about a certain point along the reference line. The rates of translation were determined from the projection of points to the reference lines. A rotation after the first translation is shown in Figure 8c. After the rotation, a translation geometry process was applied again to the actual pixels. As demonstrated by Figure 8d, the translation directions were opposite from those of the first translation. Therefore, the values of the translation were also reversed from the first translation procedure since they were vector units. The shear process played a role to decrease the error based on the reference lines, as demonstrated by Figure 8e. Finally, the dilatation process puts the actual pixels at the same positions as the target positions, as shown by Figure 8f. Equations (7)–(11) calculate the homogeneous matrices for each process. The homogeneous matrix for all processes could be determined by using the sequence of matrix multiplications denoted in Equation (12). The relationship between the actual pixel (u,v) and desired pixel (u',v') was determined by Equation (13).

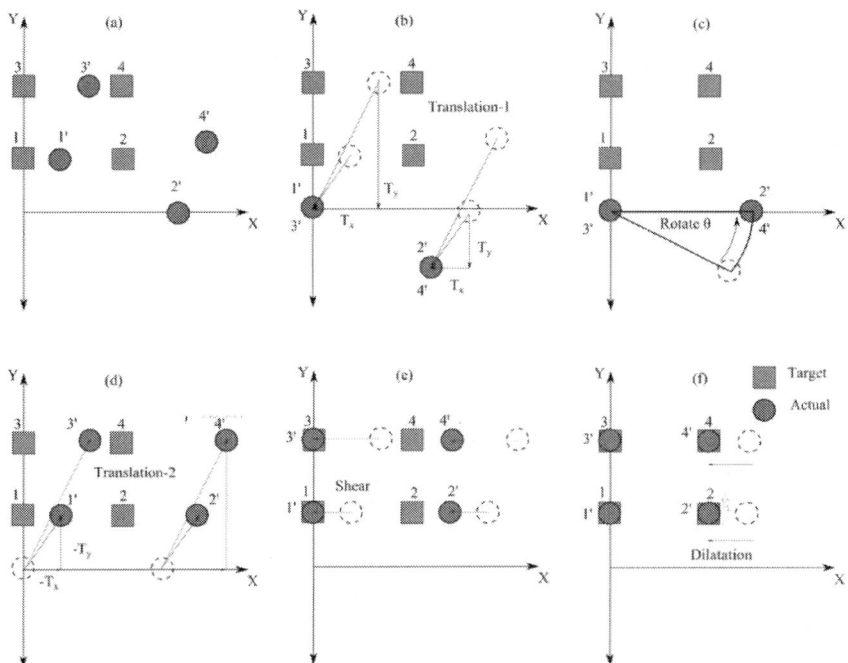

Figure 8. Geometry processes to build the eye gaze motion tracking system using the EOG. (**a**) Actual positions compared to the targets. (**b**) Translation of the reference pixel to the center of the coordinate (Translation-1). (**c**) Rotation of the actual pixel to the center of the coordinate. (**d**) Translation with opposite scale factor from the translation in (Translation-2). (b). (**e**) Shear. (**f**) Dilatation.

$$Translation-1 = \begin{bmatrix} 1 & 0 & T_x \\ 0 & 1 & T_y \\ 0 & 0 & 1 \end{bmatrix}$$

(7)

$$Rotation = \begin{bmatrix} \cos\theta & -\sin\theta & 0 \\ \sin\theta & \cos\theta & 0 \\ 0 & 0 & 1 \end{bmatrix}$$

(8)

$$Translation-2 = \begin{bmatrix} 1 & 0 & -T_x \\ 0 & 1 & -T_y \\ 0 & 0 & 1 \end{bmatrix}$$

(9)

$$Shear = \begin{bmatrix} 1 & m_1 & 0 \\ m_2 & 1 & 0 \\ 0 & 0 & 1 \end{bmatrix}$$

(10)

$$Dilatation = \begin{bmatrix} s_1 & 0 & 0 \\ 0 & s_2 & 0 \\ 0 & 0 & 1 \end{bmatrix}$$

(11)

$$Homogeneous\ Matrix = [Dilatation][Shear][Translation-2][Rotation][Translation-2]$$
$$= \begin{bmatrix} s_1(\cos\theta + m_1\sin\theta) & s_1(-\sin\theta + m_1\cos\theta) & a_1 \\ s_2(\sin\theta + m_2\cos\theta) & s_2(\cos\theta + m_2\sin\theta) & a_2 \\ 0 & 0 & 1 \end{bmatrix}$$
$$a_1 = s_1((T_x\cos\theta - T_y\sin\theta - T_x) + m_1(T_x\sin\theta + T_y\cos\theta - T_y))$$
$$a_2 = s_2((T_x\sin\theta - T_y\cos\theta - T_y) + m_2(T_x\cos\theta - T_y\sin\theta - T_x))$$

(12)

$$\begin{bmatrix} u' \\ v' \\ 1 \end{bmatrix} = Homogeneous\ Matrix \begin{bmatrix} u \\ v \\ 1 \end{bmatrix}$$

(13)

3.4. Process of Robot Control Using EOG

Figure 9 shows the overall process of controlling a robot using EOG. Gaze motions generated the EOG signals. Then, these signals were processed to get the target positions (u,v) of gaze motions based on the affine transformation in Equation (13) of actual pixels which is generated by linear relationship between EOG and gaze distance in Equation (3). Then, the pixel unita were converted to cm units by using Equation (5). The end-effector position (P_x, P_y) was calculated by Equation (4). Using inverse kinematic, the degrees of two angles were determined by Equations (A5) and (A6). These values were sent through serial communication to move the robot manipulator.

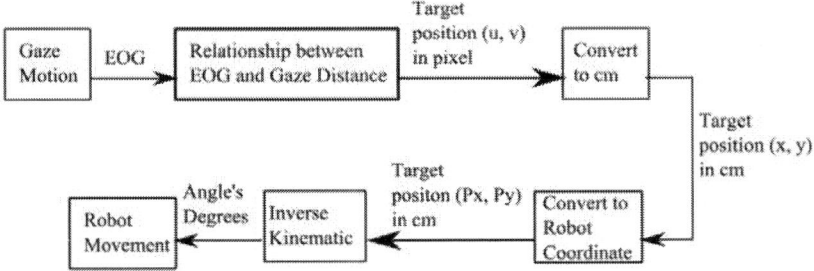

Figure 9. The process of robot control using EOG.

4. RESULT AND DISCUSSION

The tracking object system using the EOG was evaluated by three operators. The locations of the target points were exactly the same as those of the training targets, so the total number of target points was 24. Homogeneous matrices were built by the sequence of the five geometry processes. Two of the processes, which were rotation and dilatation, had rational numbers. On other hand, the translations and shear had dynamic variables. The values of the dynamic variables depended on the reference lines.

Table 3 provides the values of constant variables a and b for reference Equation (8) in the four areas. The result shows that the slopes of the reference lines followed some patterns. Areas 1 and 3 had positive slope (a) values, whereas Areas 2 and 4 had negative slope values. The highest positive gradient was 2.44 and the lowest negative gradient was −1.07. The average slopes were 1.03 in Area 1 and 1.85 in Area 3. On the other hand, the average slopes in Areas 2 and 4 were −0.20 and −0.52, respectively.

These gradients confirm that the three operators showed the same pattern for actual pixels rotating to the real target coordinates in all four areas. The EOG signals not only depended on the individual but also on the environmental condition. Therefore, fluctuations occurred for rational number b.

Table 3. Constant variables for reference equations.

Area	Variable	Operator 1	Operator 2	Operator 3
1	a	1.27	0.90	0.89
	b	6.99	−93.82	416.4
2	a	−0.30	−0.12	−0.14
	b	−155.94	68.44	11.99
3	a	2.08	2.44	1.00
	b	84.70	80.60	−241.55
4	a	−1.07	−0.41	−0.08
	b	−560.90	158.36	70.88

For the rotation process, the angle varied for all areas. Figure 10 shows the average angles for the three operators. The maximum value of the angle was 47.70° and the minimum was 13.70°. The average angle in Area 1 was 21.03°. This value was not too different compared to Areas 3 and 4, which were 25.30° and 24.60°, respectively. On the other hand, it was 36.02° in Area 2 on average, and the average of the rotation angles was 26.76°.

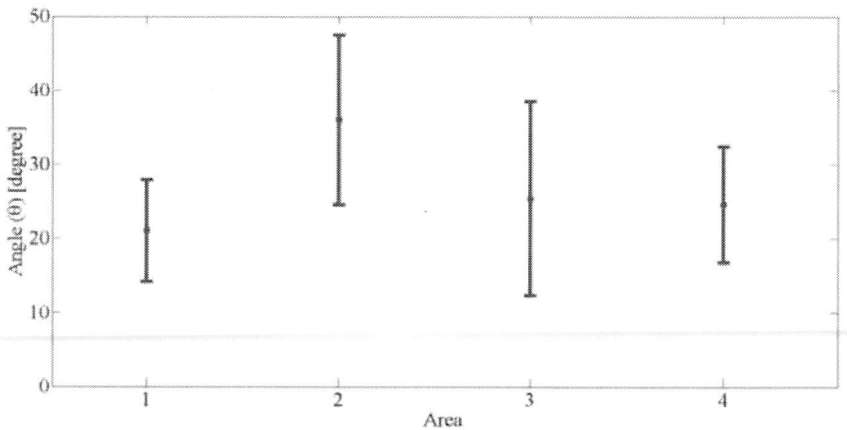

Figure 10. Average angle (θ) for the three operators for the rotation geometry process in the four areas.

Figure 11 shows the dilatation factor for the three operators in the four areas. The dilatation factors were classified into two types. The first was s_1, which reconstructed the actual pixels in the horizontal direction (x-axis). The second was s_2, which reconstructed the actual pixels in the vertical direction (y-axis). The result shows that the dilatation factor also had a pattern. For the horizontal geometry process, the average horizontal dilatation factors in Areas 1 and 3 were almost same, 0.71 and 0.63, respectively. These numbers were also the same for the vertical dilatation factor in Areas 2 and 4, which were 0.70 and 0.60. At a glance, the same values were also generated among Areas 2 and 4 in the horizontal direction and Areas 1 and 3 in the vertical direction. The average vertical dilatation factors for Areas 1 and 3 in the horizontal dilatation process were 0.93 and 0.94, whereas the average horizontal dilatation factors for Areas 2 and 4 were 0.93 and 0.92. This result shows that the actual pixels were always bigger than the target pixels. For Areas 1 and 3, the extension in the vertical direction had a bigger impact than that in the horizontal. But, Areas 2 and 4 had bigger expansion in the horizontal direction than in the vertical.

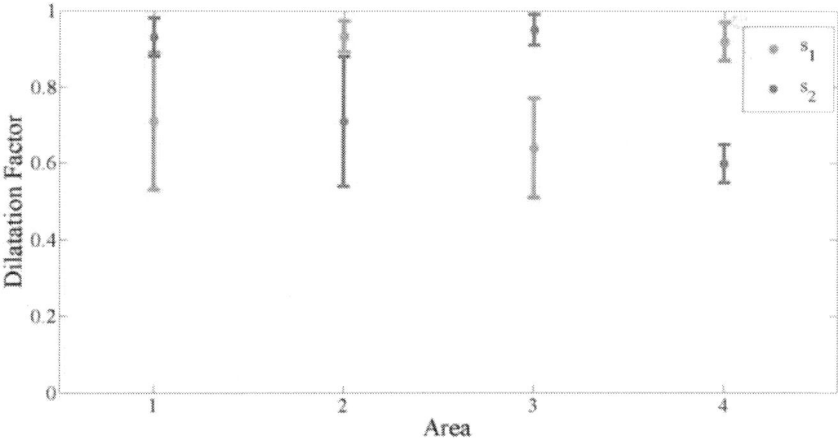

Figure 11. Average dilatation factor for horizontal (s_1) and vertical (s_2) for the three operators in the four areas.

The three operators tried to gaze at the target points. Figure 12 illustrates the actual pixel positions that the three operators gazed at while looking to the target points. The best performance of this system was when the operators gazed to the horizontal or vertical positions only. In this case, the average error in Area 1 was 18 ± 4 pixels, Area 2 was 12 ± 7, Area 3 was 11 ± 4 and Area 4 was 14 ± 9. Overall, the average error in the horizontal direction was approximately 11 ± 9 pixels and it was 7 ± 6 pixels in the vertical.

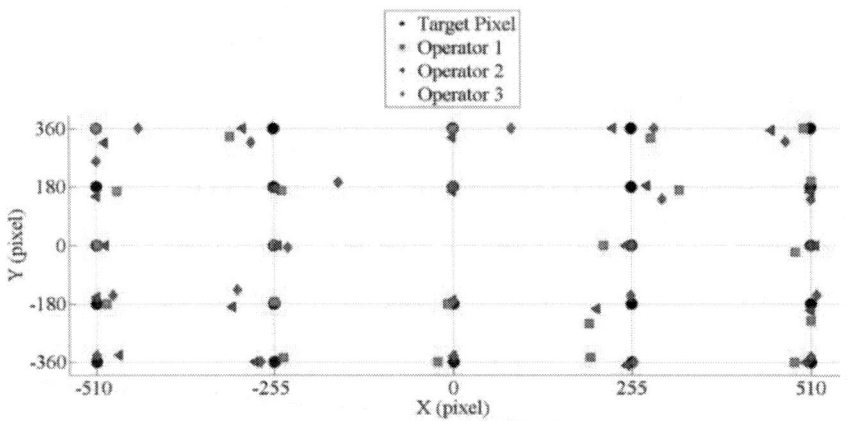

Figure 12. The average positions of actual pixels and 24 target pixels for the three operators.

The pixel errors were converted to the angle of gaze motion to make it more general. The normal distance between the operators and the targets was 40 cm. The height of target area was 27 cm and the width of it was 32 cm. Number of pixels in horizontal was 1020 pixels and number of pixel in vertical was 720 pixels. Every pixel in the horizontal direction was equal to 0.033 cm and it was 0.0375 cm in the vertical direction. So, by using trigonometry, the average error angle in the horizontal direction was 0.86° ± 0.67° and 0.54° ± 0.34° in the vertical.

The tracking system using EOG was implemented to control a robot manipulator. Table 4 shows the angle degrees of two joints, α and β, from three operators who moved the end-effector to six target points. Three operators could use the system well. They controlled the robot and it successfully tracked the gaze motions by moving it to the target points.

Table 4. Joint angles of robot manipulator for six target points from three operators.

Targets		Operator 1		Operator 2		Operator 3	
u	v	$\alpha°$	$\beta°$	$\alpha°$	$\beta°$	$\alpha°$	$\beta°$
300	180	56	107	60	105	58	109
180	−50	64	119	64	120	64	120
−300	−150	96	103	99	99	98	97
−200	180	91	82	92	81	88	84
−510	−180	111	85	111	82	110	84
510	360	52	94	52	97	54	94

5. CONCLUSIONS

This experiment was done by using a 40 cm target plate in front of an operator with limitation of the linear area to about 46° in horizontal and 38° in vertical gaze motions. Since there are many EOG recording techniques [14], electrode positions had to be first inspected in order to study the signal conditions before implementing a tracking system using EOG. The electrode positions could be different among the devices [18,19]. The skin condition is also an important factor to keep the signals stable. In this research, a low impedance and high conductivity gel was used as the surface between electrodes and skin. This system produces a stable EOG signal during the experiment as long as the head is in a fixed position.

This research improved the performance of tracking an object by using EOG signals. In other research [20], the relationship between eye movements and EOG was measured. Simulated EOG signals were identified by neural networks. This method reached a approximate accuracy of ±1.09°.

Another important issue in the rotation process was that the actual pixels were rotated to certain points that belonged to the reference lines. These lines were also useful for the translation process. The previous research [14] rotated the actual pixels to the center of the coordinate of the target pixel. The new technique proved that improvement was achieved in all areas of gaze motions.

One of the challenges of the affine transform method was obtaining a simpler method to determine the variables of the homogeneous matrices. Four homogeneous matrices were developed for the four areas, and each homogeneous matrix has the following seven variables that need to be determined: translation variables (T_x, T_y), rotation variable (θ), shear variables (m_1, m_2) and dilatation variables (s_1, s_2). This method was successfully to track the gaze motions not only in directions but also in distance that was indicated by the movement of the planar robot manipulator.

REFERENCES

1. Malmivuo, J.; Plonsey, R. Bioelectromagnetism; Oxford Unversity Press: New York, NY, USA, 1995; pp. 538–542.

2. Leigh, R.J. Electrooculogram. In Encyclopedia of the Neurological Sciences, 1st ed.; Aminoff, M., Daroff, R., Eds.; Academic Press: San Diego, CA USA, 2003; pp. 109–110.

3. Lledo, L.D.; Ubeda, A.; Ianez, A.; Azorin, J.M. Internet Browsing Application Based on Electrooculography for Disabled People. Expert Syst. Appl. **2013**, 7, 2640–2648.

4. Kumar, D.; Eric; Poole, E. Classification of EOG for Human Computer Interface. Proceedings of the 2nd Joint EMBS/BMES Conference, Houston, TX, USA, 23–26 October 2002; pp. 64–67.

5. Lv, Z.; Wu, X.; Li, M.; Zhang, D. A Novel Eye Movement Detection Algorithm for EOG Driven Human Computer Interface. Pattern Recognit. Lett. **2010**, 31, 1031–1047.

6. Barea, R.; Boquete, L.; Ortega, S.; López, E.; Rodríguez-Ascariz, J.M. EOG-Based Eye Movements Codification for Human Computer Interaction. Expert Syst. Appl. **2012**, 39, 2677–2683.

7. Guven, A.; Kara, K. Classification of Electro-Oculogram Signals Using Artificial Neural Network. Expert Syst. Appl. **2006**, 21, 199–205.

8. Banerjee, A.; Datta, S.; Pal, M.; Konar, A.; Tibarelawa, D.N.; Janarthanan, R. Classifying Electrooculogram to Detect Directional Eye Movements. Proceedia Technol. **2013**, 10, 67–75.

9. Postelnicu, C.C.; Girbacia, F.; Talaba, D. EOG-Based Visual Navigation Interface Development. Expert Syst. Appl. **2012**, 39, 10857–10866.

10. Borghetti, D.; Bruni, A.; Fabbrini, M.; Murri, L.; Sartucci, F. A Low-Cost Interface for Control of Computer Functions by Means of Eye Movements. Comput. Biol. Med. **2007**, 37, 1765–1770.

11. Itakura, N.; Sakamoto, K. A New Method for Calculating Eye Movement Displacement from AC Coupled Electro-Oculographic Signals in Head Mounted Eye-Gaze Input Interface. Biomed. Signal Process. Control **2010**, 5, 142–146.

12. Pinheiro, C.G.; Naves, E.L.M.; Pino, P.; Losson, E.; Andrade, A.O.; Bourhis, G. Alternative Communication Systems for People with Severe Motor Disabilities: A Survey. Biomed. Eng. Online **2011**, 10.

13. Deng, Y.L.; Hsu, C.L.; Lin, T.C.; Tuan, J.S.; Chang, S.H. EOG-Based Human-Computer Interface System Development. Expert Syst. Appl. **2010**, 37, 3337–3343.

14. Timmins, N.; Marmor, F.M. Studies on the Stability of the Clinical Electro-Oculogram. Doc. Ophtalmol. **1992**, 81, 163–171.

15. Rusydi, M.I.; Okamoto, T.; Sasaki, M.; Ito, S. Line of Sight Estimation from EOG Signal with Variation of Electrode Position for Human Machine Interface. Proceedings of the 6th Conference on Rehabilitation Engineering and Assistive Technology Society of Korea, Jeonju, Korea, 2–3 November 2012; pp. 234–239.

16. Rusydi, M.I.; Okamoto, T.; Sasaki, M.; Ito, S. Using EOG Signal to Operate Robot Manipulator System for Tracking Objects in 2D Space. Proceedings of International Engineering Conference, Nyeri, Kenya, 4–5 September 2013; pp. 29–35.

17. Technical Foundations of Digital Productions II. Available online: http://people.cs.clemson.edu/~dhouse/courses/401/notes/affines-matrices.pdf (accessed on 1 March 2014).

18. Hakkinen, V.; Hirvonan, K.; Hasan, J.; Kataja, M.; Varri, A.; Loula, P.; Eskola, H. The Effect of Small Differences in Electrode Position on EOG Signals: Application to Vigilance Studies. Electroencephalogr. Clin. Neurophysiol. **1993**, 86, 294–300.

19. Nojd, N.; Hyttinen, J. Modeling of EOG Electrode Position Optimization for Human-Computer Interface. Proceedings of the ICST 3rd International Conference on Body Area Networks, Brussels, Belgium, 13–15 March 2008.

20. Coughlin, M.J.; Cutmore, T.R.H.; Hine, T.J. Automated Eye Tracking System Calibration using Artificial Neural Networks. Comput. Methods Programs Biomed. **2004**, 76, 207–220.

CHAPTER 10

The Research about Prescribed Workspace for Optimal Design of 6R Robot

Yi Gan[1,2*], Weiwei Yu[1], Weiming He[1,2], Junlei Wang[1], Fujia Sun[1]

College of Mechanical Engineering, University of Shanghai for Science and Technology, Shanghai, China
Department of Precision Mechanics, Faculty of Science and Engineering, Chuo University, Tokyo, Japan

ABSTRACT

Based on the D-H notation, kinematics model and inverse kinematics model of 6R industrial robots are established. Using graphical method, the boundary curve equations of the 6R industrial robot workspace are obtained. Based on the prescribed workspace, the D-H parameter optimization method of 6R industrial robots is proposed. Using the genetic algorithm to determine the structural dimensions of a 6R robot, we make sure that its workspace can exactly contain the prescribed workspace. This method can be used to reduce the overall size of the robot, save materials and reduce the power consumption of the robot during its work time.

KEYWORDS

6R Robot, Prescribed Workspace, Optimal Design

1. INTRODUCTION

The workspace of a robot is defined as the sets of points that can be reached by the end-effecter [1] , and sometimes is known as reachable space. It is one of the main ways in the design and optimization processes of robots [2] . Based on graphical approach, Xie Jun et al. [3] analyzed the workspace of a novel series Chinese medical massage arm, and learned that the upper arm and the forearm of this kind of Chinese medical massage arm must be equal. Chen Zaili [4] presented a genetic algorithm approach for the synthesis of spatial 6-DOF parallel manipulators whose workspace must include a desired workspace with orienting capabilities on a given 3D region. M.A. Laribi [5] proposed an optimal dimensional synthesis method of the DELTA parallel robot for a prescribed workspace.

In the practical working conditions, when a robot accomplishes some special tasks, the range of motion of its end-effecter is not the whole workspace, but must be included in the whole workspace. As the range of motion usually has an irregular shape, it can be substituted by a prescribed workspace that includes this range of motion in the synthesis process of the robot [6] [7] . The prescribed workspace must satisfy two basic requirements: 1) be a regular geometric model, such as a cuboid, a cylinder or a sphere and so on; 2) must contain the entire range of motion of a robot's end-effecter and has the minimum volume.

The method which is usually used to determine the structure parameters has the property of try and blindness. It is not only a waste of time, but also hard to obtain a robot with a compact structure. Different structure parameters are different information states of every link. Aiming at obtaining a compact structure, we can determine the structure parameters which make the robot with the shortest information distance [8] . This paper, based on a prescribed workspace, presents an optimum design method of D-H parameters for a 6R robot. This method can obtain a 6R robot with a compact structure, the sum of whose links' lengths is smallest so that the material can be saved.

2. ANALYSIS FOR THE WORKSPACE OF A 6R ROBOT

2.1. D-H Coordinate Frames and Homogeneous Transformations of 6R Robot

A commonly used convention for selecting frames of reference in robotics applications is the Denavit and Hartenberg (D-H) convention which was firstly introduced by Jacques Denavit and Richard S. Hartenberg [9] . The D-H coordinate frames can be laid out as follows: 1) the Z-axis is in the direction of the joint axis; 2) the X-axis is parallel to the common normal, $X_n = Z_n \times Z_{n-1}$; 3) the Y-axis follows from the X-axis and Z-axis by choosing it to be a right-handed coordinate system. Four parameters known as D-H parameters can be obtained. They are θ, d, a, α.

The model of 6R robot is shown in Figure 1 which is used on a welding production line. The D-H coordinate frames for it are established as shown in Figure 2. And the D-H parameters are shown in Table1

Let T_j^i be the homogeneous transformation matrix from the ith D-H coordinate frame to the jth one, then the following equations can be got :

$$T_3^2 = \begin{bmatrix} c_3 & 0 & s_3 & a_3c_3 \\ s_3 & 0 & -c_3 & a_3s_3 \\ 0 & 1 & 0 & 0 \\ 0 & 0 & 0 & 1 \end{bmatrix} \quad T_2^1 = \begin{bmatrix} c_2 & -s_2 & 0 & a_2c_2 \\ s_2 & c_2 & 0 & a_2s_2 \\ 0 & 0 & 1 & 0 \\ 0 & 0 & 0 & 1 \end{bmatrix} \quad T_1^0 = \begin{bmatrix} c_1 & 0 & s_1 & a_1c_1 \\ s_1 & 0 & -c_1 & a_1s_1 \\ 0 & 1 & 0 & 0 \\ 0 & 0 & 0 & 1 \end{bmatrix}$$

$$T_6^5 = \begin{bmatrix} c_6 & -s_6 & 0 & 0 \\ s_6 & c_6 & 0 & 0 \\ 0 & 0 & 1 & 0 \\ 0 & 0 & 0 & 1 \end{bmatrix} \quad T_5^4 = \begin{bmatrix} c_5 & 0 & s_5 & 0 \\ s_5 & 0 & -c_5 & 0 \\ 0 & 1 & 0 & 0 \\ 0 & 0 & 0 & 1 \end{bmatrix} \quad T_4^3 = \begin{bmatrix} c_4 & 0 & -s_4 & 0 \\ s_4 & 0 & c_4 & 0 \\ 0 & -1 & 0 & d_4 \\ 0 & 0 & 0 & 1 \end{bmatrix}$$

where, $c_i = \cos(\theta_i),\ s_i = \sin(\theta_i)\ i = 1, 2, \cdots, 6$.

The transformation matrix from the base to the end-effecter is given as follows:

$$T_6^0 = T_1^0 T_2^1 T_3^2 T_4^3 T_5^4 T_6^5 = \begin{bmatrix} n_x & o_x & a_x & p_x \\ n_y & o_y & a_y & p_y \\ n_z & o_z & a_z & p_z \\ 0 & 0 & 0 & 1 \end{bmatrix}$$

(1)

2.2. Analysis and Simulation of the Workspace

Twelve messages are given by the matrix with Equation (1). n_x, n_y, n_z, o_x, o_y, o_z, a_x, a_y and a_z are nine pose information that determine the posture of the robot, p_x, p_y and p_z are three position information that determine the position of the robot. As the scope of the workspace of a 6R robot is determined by position information, so we can just calculate p_x, p_y and p_z when we analysis the workspace. p_x, p_y and p_z are given with Equation (2).

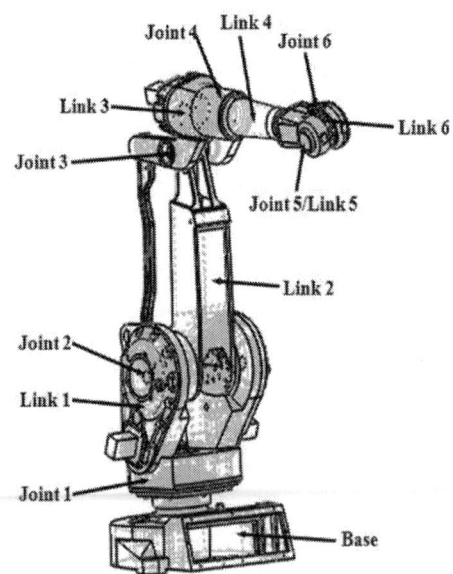

Figure 1. The three-dimensional model of the 6R robot.

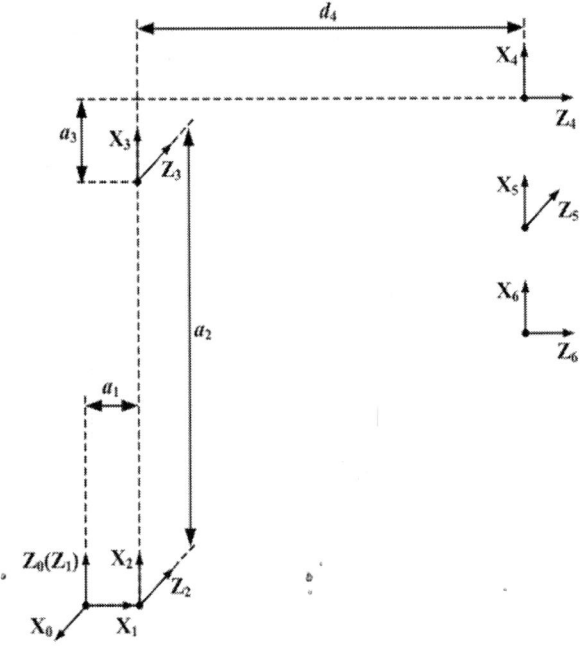

Figure 2. The D-H coordinate frames of the 6R robot.

Table 1. The D-H parameters of the 6R robot.

Joint	θ_i (°)	d_i (°)	a_i (mm)	α_i (mm)	Range of θ_i (°)
1	θ_1	0	$a_1 = 100$	90	−180 ~ +180
2	θ_2	0	$a_2 = 705$	0	−100 ~ +110
3	θ_3	0	$a_3 = 135$	90	−60 ~ +65
4	θ_4	$d_4 = 755$	0	−90	−200 ~ +200
5	θ_5	0	0	90	−120 ~ +120
6	θ_6	0	0	0	−400 ~ +400

$$\begin{cases} P_x = c_1 \left(a_1 + a_2 c_2 + a_3 c_{23} + d_4 s_{23} \right) \\ P_y = s_1 \left(a_1 + a_2 c_2 + a_3 c_{23} + d_4 s_{23} \right) \\ pz_x = a_2 s_2 + a_3 c_{23} - d_4 c_{23} \end{cases}$$

$$(2)$$

where $c_{23} = \cos(\theta_2 + \theta_3)$, $s_{23} = \sin(\theta_2 + \theta_3)$. From Equation (2), we can infer that the position of the end-effecter of a 6R robot is only defined by the first three joints (Joint 1, Joint 2 and Joint 3 in Figure 1). So we can define the end of the third link (the fourth joint point) as the reference work point of the 6R robot, and the set of points which the reference work point can arrive at can be defined as the workspace of the 6R robot [10].

Let $\theta_1 = 0°$, according to the graphical method, the cross-section of the workspace of the 6R robot on the plane $X_0O_0Z_0$ can be obtained [11]. It is the area which is surrounded by the arc line Γ_1, Γ_2, Γ_3 and Γ_4 in Figure 3. The workspace of the 6R robot can be obtained by rotating the cross-section around the axis O_0Z_0. The three-dimensional view and the three-dimensional cross-sectional view of the workspace are respectively shown in the Figure 4 and Figure 5. From the Figure 4, we can see that the workspace is similar to a sphere. From the figure 5, we can see that there is a cavity in the workspace. When designing 6R robots, we must make sure that the cavity doesn't intersect with the reference work point of 6R robots.

3. 6R ROBOT D-H PARAMETERS OPTIMIZING WITH PRESCRIBED WORKSPACE

3.1. Object function of Robot Structure Optimizing Based on Prescribed Workspace

According to the given n end-effecter working points of robot, the minimum rectangular can be determined which includes these n working points [6]. As shown in the Figure 6, the prescribed workspace of Δ_1 is the minimum rectangular which includes all of the working points of 6R. The size of Δ_1 is $L_1 \times L_2 \times L_3$. Δ_1 is symmetry relative to $X_0O_0Y_0$ plane and $Y_0O_0Z_0$ plane. Figure 7 shows the position of Δ_1 and the workspace.

The optimization objective function of Robot link length is as follows,

$$\mathrm{Min}F(X)=l_1F_1(X)+l_2F_2(X) \quad X \in R^+$$

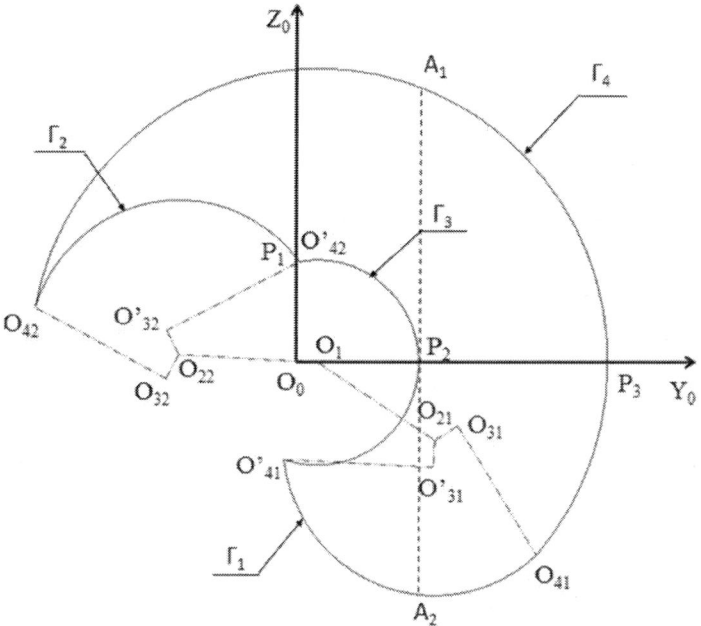

Figure 3. The section shape of the workspace of 6R robot on the $XOOOZ$ plane.

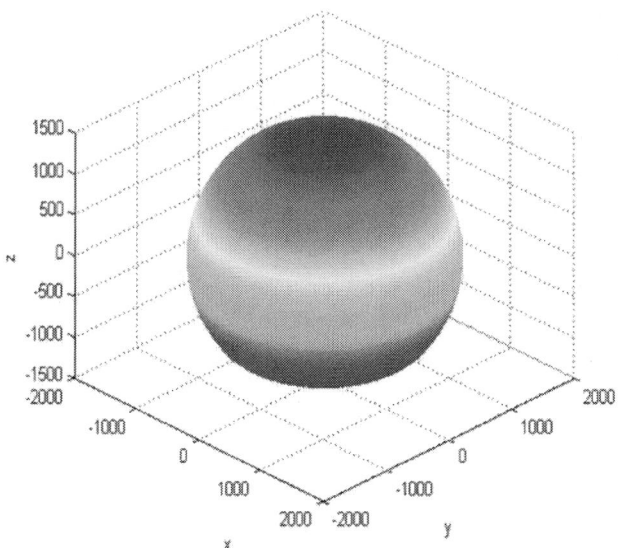

Figure 4. The three-dimensional view of the workspace.

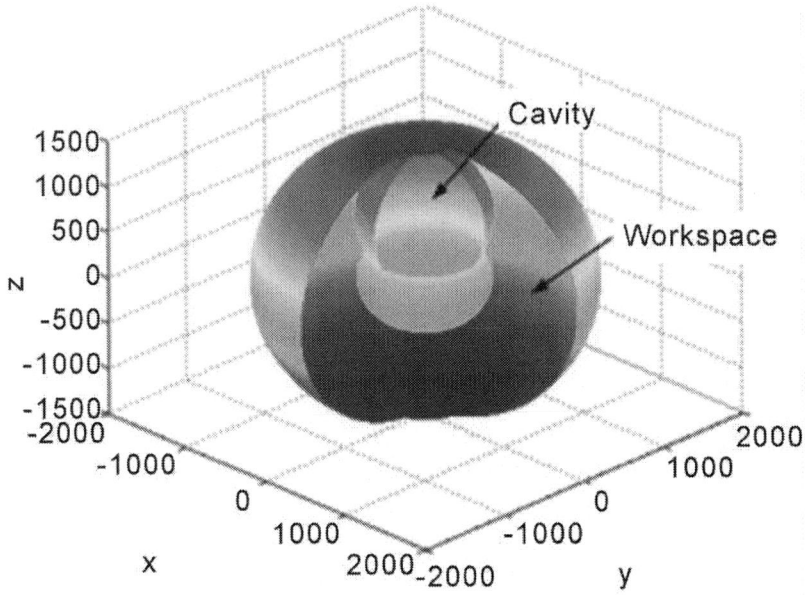

Figure 5. The three-dimensional cross-section view of the workspace.

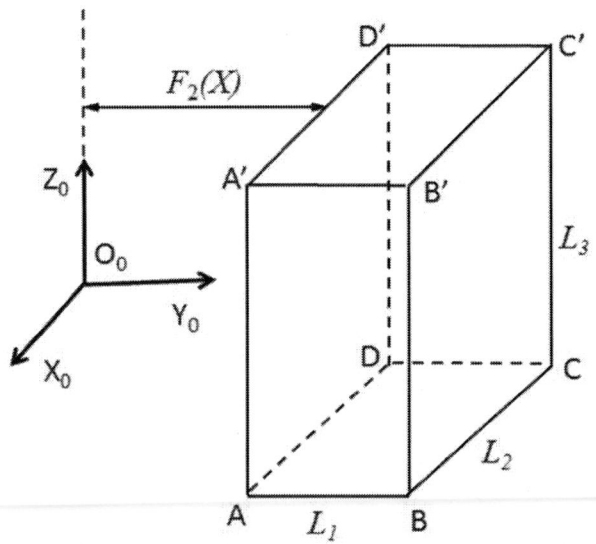

Figure 6. The prescribed workspace.

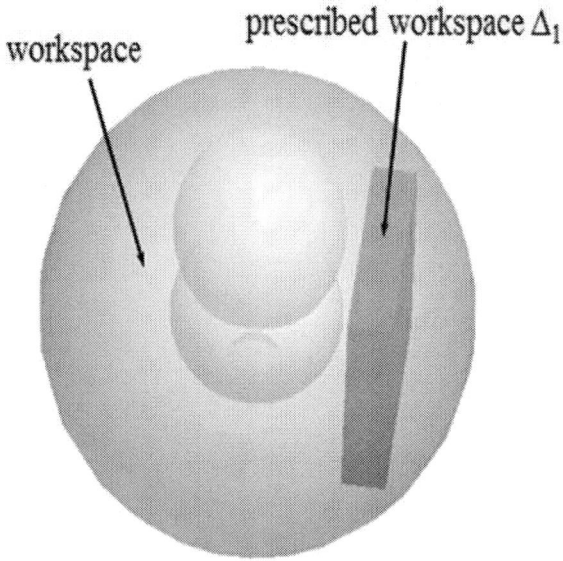

Figure 7. The prescribed workspace and the workspace.

And the constraint conditions are,

$$\begin{cases} f_k(Y) \le 0 & (k = 1,2,\cdots,t) \\ 0 < x_{i\min} \le x_i & (i = 1,2,\cdots,m) \\ q_{j\min} \le q_j \le q_{j\max} & (j = 1,2,\cdots,n) \end{cases}$$

λ_1, λ_2 are constants, $\lambda_1, \lambda_2 \hat{I}(0,1)$, and $\lambda_1 + \lambda_2 = 1$. $F_1(X)$ is sum of links length. $F_2(X)$ is the distance between the prescribed workspace and the origin O_0 of the coordinate $O_cX_0Y_cZ_c$. x_i is the length of ith link, m is the number of the links, q_j is the jth joint angle, n is the number of the joints. X is m dimensional vector, $X = [x_i]^T$. Y is m + n dimensional vector, $Y = [X,Q]^T$.

Because the positions of 6R robot end-effecters are affected by angles of the first three joints (Joint 1, Joint 2 and Joint 3 in Figure 1), the length of the first four links (Link 1, Link 2, Link 3 and Link 4 in Figure 1) can be optimized based on the prescribed workspace. The object function is :

$$\mathrm{Min}F(X) = \lambda_1 (x_1 + x_2 + x_3 + x_4) + \lambda_2 D(x_1, x_2, x_3, x_4) \quad (3)$$

$D(x_1, x_2, x_3, x_4)$ is the distance between the plane of ADD'A' in Figure 6 of prescribed workspace and the origin O_0 of the coordinate $O_0 X_0 Y_0 Z_0$ in Figure 6. The value of $D(x_1, x_2, x_3, x_4)$ is determined by the size of the internal cavity of work space.

The angle between the first axis of link 1 in Figure 1 and the horizontal plane is γ. According to the rules of D-H coordinate system, $a_1/\cos(\gamma) = x_1$, $a_2 = x_2$, $a_3 = x_3$ and $d_4 = x_4$. And the object function can be deformed as follows, then the MinF(X) is equal to Equation (4), that is

$$\mathrm{Min}\{\lambda_1(x_1 + x_2 + x_3 + x_4) + \lambda_2 D(x_1, x_2, x_3, x_4)\} \quad (4)$$

$$\mathrm{Min}F(X) = \lambda_1(x_1 + x_2 + x_3 + x_4) + \lambda_2 D(x_1, x_2, x_3, x_4)$$
$$= \lambda_1(a_1/\cos(\gamma) + a_2 + a_3 + d_4) + \lambda_2 D(a_1, a_2, a_3, d_4)$$

3.2. Constraint Conditions

3.2.1. Length of the Links Constraining

Considering the actual work requirements, the length of the links can't be small too much. The ranges of a_1, a_2, a_3 and d_4 are satisfied the following rules.

$$a_1 \geq l_1; a_2 \geq l_2; a_3 \geq l_3; d_4 \geq l_4 \quad (5)$$

3.2.2. Joint Angles Constraining

In Figure 3, the length of links and the joint angles were certain size, the point of O_{42} is in the extended line of O_1 and O_2 when the section area of workspace is the biggest. At the same time, $\theta_{3\,\mathrm{max}} = \arctan(d_4/a_3)$.

Considering the actual situations and the interference problems of the structure, the rotation range of the joints should be limited. The ranges of θ_2 and θ_3 can be set as follows.

$$\theta_{2\min} \le \theta_2 \le \theta_{2\max}; \theta_{3\min} \le \theta_3 \le \arctan\left(d_4/a_3\right)$$

(6)

3.2.3. Internal Cavity Constraining

In Figure 3, let M_1 be the radius of Γ_2 and Γ_3, let M_2 be the radius of Γ_4. Then

$$M_1 = \sqrt{a_3^2 + d_4^2}$$

$$M_2 = \sqrt{a_2^2 + a_3^2 + d_4^2 + 2a_2 M_1 \cos\left(\theta_{2\min} - \varphi_1\right)}$$

where $\varphi_1 = 2\pi + \theta_{2\min} + \theta_{3\min} - \theta_{2\max}$.

If the sum of a_1, a_2, a_3 and d_4 is the smallest, then the shadow of Δ_1 on the plane of $Y_cO_0Z_0$ should be on the right to the connection of A_1 and A_2 in Figure 3. To ensure the cavity do not cross Δ_1, the follows should be met.

$$D\left(a_1, a_2, a_3, d_4\right) = \max\left(x_{p1}, x_{p2}\right)$$

(7)

where,

$$x_{p2} = \begin{cases} a_1 + M_2 & (\gamma_2 > 0) \\ a_1 + M_2 \cos\left(\gamma_2\right) & (\gamma_2 \le 0) \end{cases}$$

$$x_{p1} = \begin{cases} a_1 + a_2 \cos\left(\theta_{2\max}\right) + M_1 & (\gamma_1 \le 0) \\ a_1 + a_2 \cos\left(\theta_{2\max}\right) + M_1 \cos\left(\gamma_1\right) & (\gamma_1 > 0) \end{cases}$$

γ_1 is the included angle of the line of O'_{42} and O_{22} and the line of O_0Y_0 in Figure 3. $\gamma_1 = \theta_{2\max} + \theta_{3\min} - \theta_{3\max}$. γ_2

is the included angle of the line of O'_{42} and O_1 and the line of O_0Y_0 in Figure 3. $\gamma_2 = \theta_{2\max} - \varphi_2$. φ_2 is the included angle of the line of O'_{42} and O_1 and the line of O_1 and O_{22} in Figure 3.

3.2.4. Key Points Constraining

Let the y coordinate of O_{41} less than or equal to the value of $D(x_1, x_2, x_3, x_4)$ with the internal cavity constraining. Then the section which is on the right of the connection line of A_1 and A_2 is symmetry relative to the O_0Y_0 axis. The space of Δ_2 can be got to revolve the section around the O_0Z_0 axis. Δ_2 is symmetry relative to $X_cO_0Y_0$ plane and $Y_0O_0Z_0$ plane. If the point of $C'(x'_c, y'_c, z'_c)$ is in Δ_2, then Δ_1 must be included in Δ_2 and in the workspace of 6R robot. And the points constraining is as follows.

$$x_{o41} \le D(a_1, a_2, a_3, d_4) \tag{8}$$

$$\begin{cases} D(a_1, a_2, a_3, d_4)) + L_1 \le a_1 + M_3 \cos(\gamma_4)\cos(\gamma_3) \\ L_2/2 \le a_1 + M_3 \cos(\gamma_4)\sin(\gamma_3) \\ L_3/2 \le M_3 \sin(\gamma_4) \end{cases} \tag{9}$$

where, $M_3 = a_2 + M_1$; $\gamma_3 = \arctan\left[(l_2/2)/(l_1 + a_1 + M_2)\right]$;

$\gamma_4 = \arccos\left\{\left[(l_2/2)/\sin(\gamma_3) - a_1\right]/M_2\right\}$.

4. APPLICATION

According to the location and distribution of a car body welding, the space sizes of prescribed workspace of one 6R robot are

$$L_1 \times L_2 \times L_3 = 500 \text{ mm} \times 1650 \text{ mm} \times 1400 \text{ mm} \tag{10}$$

In order to avoid the movement interference, the links length and the joints angle robot are limited. $l_1 = 100$ mm, $l_2 = l_3 = l_4 = 150$ mm, $-120° \leq \theta_2 \leq +120°$, $-70° \leq \theta_3$. Let $\lambda_1 = 0.65$, $\lambda_2 = 0.35$ in 3th equation. Based on the genetic algorithm [12] , the optimal solution of the constraint conditions are shown in Table2 And the distance between the prescribed workspace to the plane of $O_CX_0Y_CZ_0$ is $D(a_1,a_2,a_3,d_4) = 483.94$ mm.

According to the optimal solution, the workspace can be got by the reference [13] . Δ_1 was created by the methods above mentioned showing Figure 8. And Δ_1 was cut respectively by the surfaces of A, B, C showing from Figure 9 to Figure 12. Δ_1 was just included in the workspace and Δ_1 doesn't intersect with the internal cavity. It proves the rationality of optimization result.

Rounding the optimized size in Table 2, $a_1 = 100$ mm, $a_2 = 670$ mm, $a_3 = 150$ mm, $d_4 = 690$ mm. Then the actual robot can be created with the optimized size showing in Figure 13.

In market, the workspace of robot of IRB 2400/10 can meet the 10th equation. The contrast of these two kinds of robots shows in Table3

Table 2. The optimal solution.

Variables	a_1 (mm)	a_2 (mm)	a_3 (mm)	d_4 (mm)	$\theta_{2\,min}$ (°)	$\theta_{2\,max}$ (°)	$\theta_{3\,min}$ (°)
Results	100.2	669.6	150.9	689.9	−104.8	120	−70

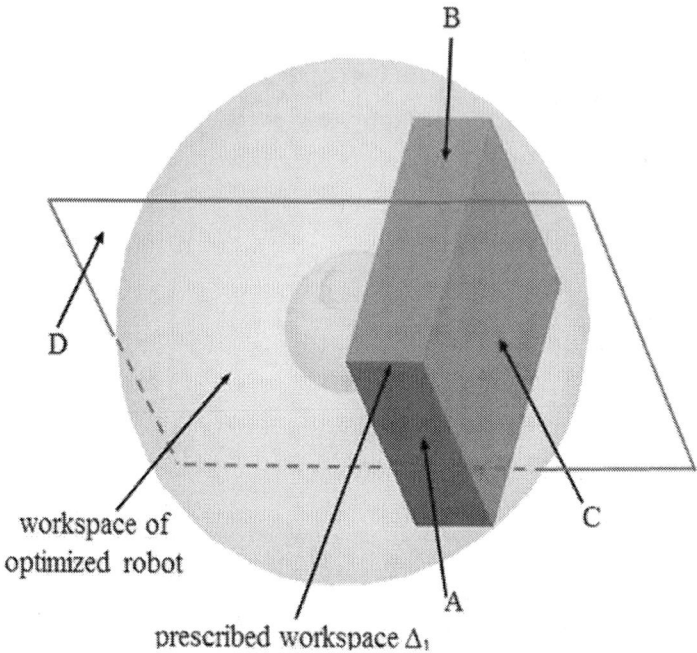

B

D

workspace of
optimized robot

prescribed workspace Δ_1

C

A

Figure 8. The prescribed workspace and the workspace of optimized robot.

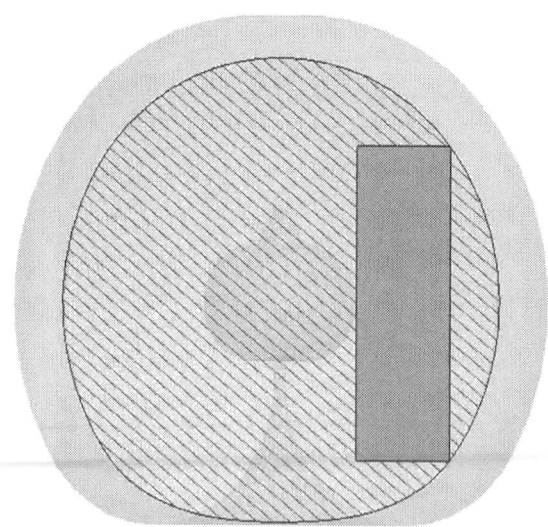

Figure 9. The section of the optimized workspace on surface A.

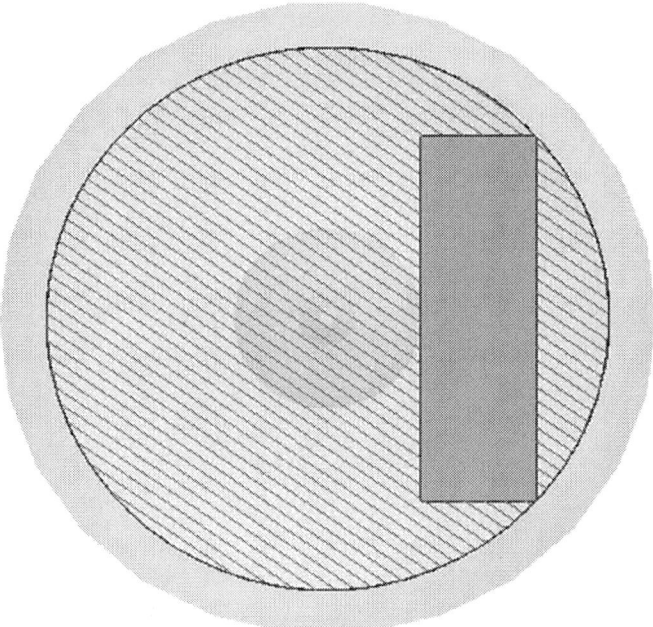

Figure 10. The section of the optimized workspace on surface B.

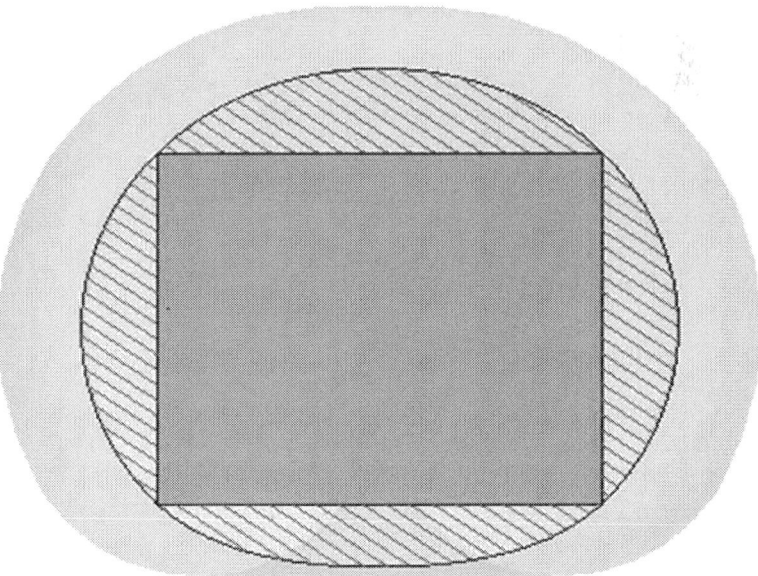

Figure 11. The section of the optimized workspace on surface C.

In Table 3, the 2 kinds of workspace volume are small difference. The sum of link length of the optimized robot is shorter than that of IRB 2400/10 robot by 85 mm. And the weight of the former is 7.89% lighter than that of the latter.

5. CONCLUSION

It analyzes the workspace of 6R robot to ensure the joints that affect 6R robot's work space. And the edge curve of work space had been got by the graphic method. Matlab was used to establish the simulation model of 6R robot work space.

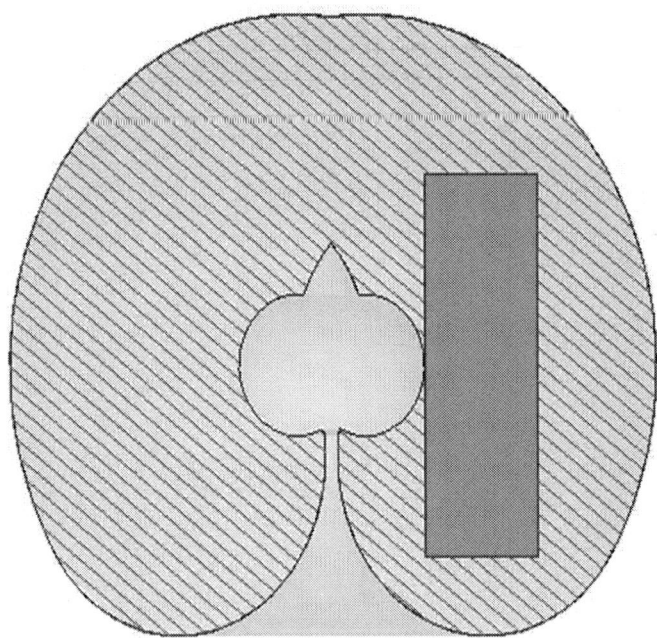

Figure 12. The section of the optimized workspace on surface D.

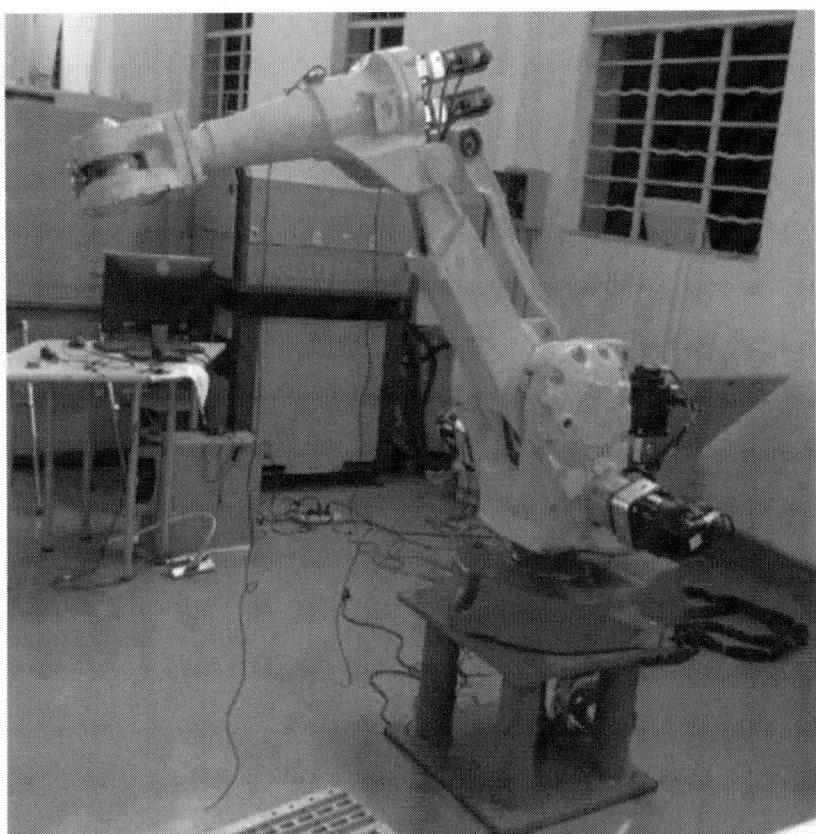

Figure 13. The optimized robot.

Table 3. The optimal solution.

	2nd link length (mm)	3rd link length (mm)	4th link length (mm)	Workspace volume (m³)	Weight (Kg)
Optimized robot	670	150	690	12.15	350
IRB 2400/10	705	135	755	13.6	380
Change	−35	+15	−65	−1.54	30

With the prescribed workspace, the D-H parameters were optimized with GA to achieve the optimized solution meeting the constraining. The workspace and the prescribed workspace were modeled with Pro/E to prove the rationality of optimization result.

ACKNOWLEDGMENT

Project was supported by the National Natural Science Foundation of China (No. 51375314).

REFERENCES

1. Alciatore, D. and Ng, C. (1994) Determining Manipulator Workspace Boundaries Using the Monte Carlo Method and Least Squares Segmentation. ASME Robotics: Kinematics, Dynamics and Control, 72, 141-146.

2. Zhang, L.J., Niu, Y.W. and Li, Y.Q. (2009) Research on Workspace of a Spherical 2-DOF Parallel Manipulator with Actuation Redundancy. China Mechanical Engineering, 20, 2974-2978.

3. Xie, J., Kuang, L.H. and Ma, L.Z. (2011) Type Synthesis and Analysis of Workspace of a Novel Series Chinese Medical Massage Arm. China Mechanical Engineering, 22, 697-701.

4. Chen, Z.L., Chen, X.S. and Xie, T. (2002) The Synthesis of Spatial Parallel Manipulators for a Specific Workspace with a Genetic Algorithm. China Mechanical Engineering, 13, 187-190.

5. Laribi, M.A., Romdhane, L. and Zeghloul, S. (2007) Analysis and Dimensional Synthesis of the DELTA Robot for a Prescribed Workspace. Mechanism and Machine Theory, 42, 859-870.

6. Bi, Z.M., Wu, R.M. and Cai, H.G. (1994) Workspace Synthesis of Industrial Robot. Robot, 16, 181-184.

7. Chablat, D., Moroz, G., Arakelian, V., et al. (2012) Solution Regions in the Parameter Space of a 3-RRR Decoupled Robot for a Prescribed Workspace. In: Latest Advances in Robot Kinematics, Springer Netherlands.

8. Yi, G. and Wang, J.L. (2013) Studies on Information States Measurement for Modeling Design. Applied Mathematics Information Sciences, 7, 627-632.

9. Zhan, X.L., Xin, H.B. and Lente, H.-P. (2010) Kinematics Simulation of MOTOMAN-HP3 Robot Based on Virtual Reality. China Mechanical Engineering, 21, 1952-1954.

10. Zhang, P.C. and Zhang, Y. (2010) Study on Workspace Analysis of 6R Robot Based on Envelope Method. Machinery Design & Manufacture, 10, 164-166.

11. Duan, Q.J., Huang, D.G. and Li S.N. (1996) The Graphic Methods of Robot Workspace and Inscribed Cube. Journal of Nan Jing University of Science and Technology, 20, 318-321.

12. Xiao, Z.Q. and Cui, L.L. (2006) GA Based Concurrent Optimization and Design of Flexible Manipulator System. Robot, 26, 170-175.

13. Xie, F., Chen, L.M. and Zhang, C.Y. (2008) Solution Joint Robot Workspace Based on Pro/E. Machine Tool & Hydraulics, 36, 145-146.

CHAPTER 11

Deliberation on Design Strategies of Automatic Harvesting Systems: A Survey

Shivaji Bachche

Institute of Multidisciplinary Research for Advanced Materials, Tohoku University, Katahira 2-1-1, Aoba-ku, Sendai, Miyagi 980-8577, Japan

Abstract

In Asia, decreasing farmer and labor populations due to various factors is a serious problem that leads to increases in labor costs, higher harvesting input energy consumption and less resource utilization. To solve these problems, researchers are engaged in providing long term and low-tech alternatives in terms of mechanization and automation of agriculture by way of efficient, low cost and easy to use solutions. This paper reviews various design strategies in recognition and picking systems, as well as developments in fruit harvesting robots during the past 30 years in several countries. The main objectives of this paper are to gather all information on fruit harvesting robots; focus on the technical developments so far achieved in picking devices; highlight the problems still to be solved; and discuss the future prospects of fruit harvesting robots.

KEYWORDS

fruit harvesting robots; greenhouse robots; recognition system; multispectral imaging; fruit picking robots; fruit harvesting manipulator; robot; end-effectors

1. INTRODUCTION

Agriculture and food are the backbone of many developed and under developing countries that helps countries to improve their economic, social and individual status. Agriculture is also one of the main reasons to bring humans together resulting in the establishment and development of human civilizations around the globe over the past 10,000 years. The high-tech, precise and qualitative large-scale modern agriculture industry of today is a result of evolutions in time and different inventions in agriculture. The present era of modern high-tech and controlled environmental agriculture is producing good quality food, taking care to meet the basic nutritional needs for human health. The major changes in agriculture have occurred through domestication of crops and animals, weed control techniques, water management, fertilizer/pesticide application, genetic engineering and the large scale mechanization that ensued in the mid-1990s. These major changes helped the agriculture sector to grow up rapidly with mechanization and precision technologies by discovering incredible innovations and bringing on various revolutions around the world.

In recent decades, advanced technology and the latest results of scientific research have been largely applied in agriculture in order to improve the quality of products and to increase productivity. The rapid growth in the world population demands a constant quality food supply. In Asia, decreasing farmer and agricultural labor populations due to various factors is a serious problem, especially in Japan [1]. As a result, to solve this problem, researchers are engaged in providing long term and low-tech solutions in terms of mechanization and automation of the agriculture sector by using highly sophisticated robots that can replace manpower, in tasks where a person may perform worse than an automatic device in terms of precision, consistency and working cycle. The application of automation in greenhouses is very common these days; especially, modern high-tech greenhouses are equipped with automatic machines and control systems which are derived versions of numerically controlled machines.

Fruit harvesting is an important application in greenhouse horticulture that helps to save on labor costs and harvesting energy consumption, and to improve resource utilization [2,3,4]. In agriculture, some damage resistant agricultural products like olives and almonds can be harvested using trunk or branch shakers [5]. However, delicate fruits, such as tomatoes, oranges, apples or strawberries, for fresh markets cannot be harvested using aggressive methods like shakers. If these methods were used, the fruits could be damaged by being impacted by the branches of the tree during the fall or by the tree directly falling on the ground, and therefore fruit would lose quality and would this result in a reduction of trading income from the fresh produce market. Also, there is the chance of detaching unripened or small, immature fruits by shaking the trunk or branches of a tree [6]. Again, manpower will be required to collect the fruits dropped on the ground after shaking, resulting in increased labor and harvesting operation costs.

On the other hand, the manual fruit harvesting method is highly labor intensive and inefficient in terms of both economy and time. To perform intensive manual harvesting, large labor power is required and at the same time labor wages are constantly rising. The only way to maintain or reduce labor costs per unit of output is to increase productivity of labor or increase the volume of output. Competing on low labor costs is infeasible, given world trade laws and costs of living. Hence, mechanization is the only answer, since it offers, potentially, the only option for reducing harvesting labor expenses, so that growers can stay competitive in the years ahead and even markets can expand [7]. Also, mechanization plays a vital role in securing the future of fruit growers in developed countries. Moreover, in addition to providing means for reducing the drudgery of harvest labor and as the only solution to maintaining harvest productivity, harvest machinery improves the ability of farmers to perform operations in a timely manner. It also reduces the risks associated with the need for large amounts of seasonal hand labor for short periods of time and lessens the social problems which accompany an excessive influx of low-wage workers. The machine harvesting systems are a partial solution to overcome these issues by removing fruits from the trees efficiently; thus, reducing the harvesting cost to about 35%–45% of the total production cost [2] and helping to save the labor cost and harvesting energy consumption and improving resource utilization in agricultural activities [8]. By considering the above mentioned issues and the necessity for, and potential of, fruit harvesting robotics in agriculture, this review paper was presented to provide all the required data and developments that took place in the last 30 years while focused on a single issue. Most of the data regarding fruit harvesting robotics are scattered or dispersed throughout the scientific and technical journals; this paper is key to bringing all the information together in one paper and as a basis for novice researchers to build on. Although there are several review papers available, this paper is in an updated form that can provide an insight to developments in fruit harvesting robotics over three decades throughout the world. This descriptive paper addresses various design strategies in recognition and picking systems, and developments in fruit harvesting robots in the past 30 years to address the above mentioned issues. Section 2 represents the design strategies for picking and recognition systems. Section 3 gives an insight into the developments that took place in fruit harvesting robotics over the last three decades in chronological order while Section 4 discusses the present challenges and future prospect of fruit harvesting robots as a commercial product.

2. DESIGN STRATEGIES

Fruit harvesting robots usually consist of three main units; the first unit is a recognition system in which identification and location of fruits are confirmed, the second unit is a picking system in which grasping and cutting operations are performed; and the third unit is a moving system in which programmed based sub-unit of the robot moves inside the farm or in the furrows during a harvesting operation in greenhouses. Depending on the agricultural application and on the workspace in which the robot will operate, a rotational joint, linear joint, twisting joint, revolving joint and orthogonal joint or a combination of these joints are used to connect the links which form a revolute, spherical, cylindrical, rectangular or telescopic robot structure. The links are further equipped with actuators such as hydraulic pumps, air cylinders, linear actuators or electric motors for output motion. Mechanical components such as gears, bearings, belts or linkages are used to transform output motion from actuators. The feedback sensors such as optical encoders, resolvers, thermocouples, cameras or motion detectors are used to measure the various parameters and provide feedback to the control unit. A motion controller used to generate set points for providing reference measurements while a drive or amplifier is used to transform the control signals. The motion of the robot can be controlled by several control functions such as velocity control, position control, pressure or force control, and electronic gearing, and every function consists of several methods or operations mechanisms to perform the operation. Robotic operations and movements can be controlled by sequential looping programming through the computer. Now-a-days, a number of commercial software programs are available to study the dynamic and kinematic behavior of the robots and obtain motion trajectories.

In recognition systems, cameras such as CCD camera, infrared camera, high-speed camera or multispectral camera are used along with artificial lighting systems if required. A captured image from a camera is then transferred to the computer and a specific image processing algorithm based on particular feature attributes and color specification models are adopted to process the images captured by cameras. The image processing system discriminates the fruit/fruits in natural background and provides three-dimensional location (X, Y, Z coordinates) and orientation of fruit i.e., length of fruit stem and vertical angle of stem. The location helps to move the picking unit to target the fruit while orientation helps to determine grasping and cutting points. The information used to detect and locate the fruits on trees includes shape, size, edges or color while methods and algorithms used to discriminate fruit changes with physical, chemical or geometrical properties of fruits [9,10,11,12]. Also, there are numerous approaches for image processing and data analysis used in recognizing fruits, which shows the importance of fruit recognition systems in harvesting robotics [13]. Figure 1 represents a conceptual design of a fruit harvesting robot.

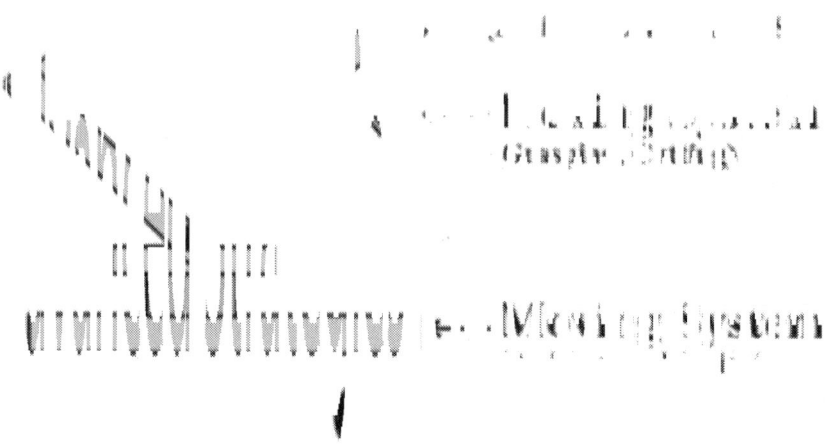

Figure 1. Overview of a fruit harvesting robot.

2.1. Picking System

The robot grippers in horticulture applications for fresh fruit and vegetable manipulation have to fulfill special requirements such as high speed activation, adaptation to a variety of shapes, maximum adherence and minimal pressure, no damage to the product, low maintenance, high reliability, low weight, be approved for contact with foodstuffs, low energy consumption, required positional precision for both gripping and releasing of the product, ease of cleaning, and easy and fast ejection of the product (important for products of low weight). By considering these special requirements and to specify gripping manipulation, Blanes et al. [14] classified the manipulation strategies to design grippers based on the above mentioned factors.

For fruit harvesting robots, the direct contact type strategy using pneumatic, hydraulic and electrical methods is always efficient, as there is less damage to fruit. To reduce the mechanical damage to the fruit during harvesting, several grasp theories and stability analysis were presented

[15,16,17,18,19,20,21,22,23,24,25] by considering the curvature of both the fingers and the object at multi-contact points, and the effect of curvature and stiffness on the stability of grasps with and without friction at the point contact was investigated. According to these grasp theories, to facilitate control, a simple end-effector with two parallel curved type fingers always shows better and steady grasping stability than plate fingers. Table 1 provides comparison of electrical, pneumatic and hydraulic gripping systems which helps to determine the specific strategy for particular application. In agriculture, electric and pneumatic grippers have shown good results for single and cluster fruits with high accuracy, repeatability, easy maintenance and small size which make them popular for fruit harvesting robots these days.

Table 1. Comparison of motors manipulation.

	Electric Grippers	Pneumatic Grippers	Hydraulic Grippers
Accuracy, strength and speed	High accuracy and repeatability, good strength, high speed	High accuracy, good strength, high speed	Good accuracy, high strength, high speed
Space	Less floor space	Less floor space	Large floor space
Advantage	Low cost and easy maintenance	Easy maintenance	Mechanical simplicity
Disadvantage	Easy to damage	Needs precise system control	Used usually for heavy payloads
Application	Good results for single fruits	Good results for cluster fruits	Good results for single fruits

After determining the gripping and manipulation strategies, it is always important to determine the interactive factors and design parameters relevant to the strategies and desired gripping application. All the possible interactive factors and input parameters relevant to gripper design need to be framed together which provides the steps to obtain the optimal gripper design. The factors that significantly influence the gripper selection were given by Monkman, et al. [26], such as the robot and machine system, position of components in installation/equipment, size and shape, mass and properties, motion sequence, velocity and acceleration, forces acting on gripper during motion, grasping points, gripper drive parameters, etc., in which the crucial conditions for not only dynamic but also static components are interconnected with the optimal gripper design. After deciding on the important parameters and primitive design, it is always good to perform modeling and simulation of the design in appropriate 3D mechanical CAD software to verify the specified parameters and performance under a controlled environment. Edan et al. [27] reported a finite element modeling technique and optimization of modeling parameters for melon grippers while Bachche et al. [28] and Bachche and Oka [29] described the simulation and modeling methods using Solidworks for a sweet pepper harvesting robot hand (Figure 2). The modeling of the system provides all optimal parameters which can be used to build the prototype and improve on the system performance obtained in the simulation process. The simulation process helps to determine the static and fatigue characteristics and effectiveness of the system. In case of any failure within the system, a part of the system can be redesigned or material can be changed to obtain highly stable and efficient end-effector before the actual manufacturing process takes place. These techniques and methods always help to optimize the picking system parameters through several design studies and it is also possible to perform the motion analysis of the model without actually prototyping it. The final optimized parameters which interact with the environment can be used for prototyping which always saves time, manufacturing costs and complicated calculations. The static characteristics such as stress, strain and displacement analysis help to determine the performance of the system under a working environment while fatigue characteristics such as fatigue life cycle and fatigue load factor plots helps in determining the component analysis for better performance of the designed system.

Monta et al. [30] developed two types of end-effectors for tomato harvesting robots based on the physical properties of tomato. First end-effector prototype has two parallel plate fingers and a suction pad while the second prototype has air pressure pads replacing the suction cup. First prototype is unable to harvest fruits with a short peduncle while the second prototype could harvest fruits regardless of peduncle length. Sakai et al. [31] provided designs based on parallel type manipulation for heavy material handling manipulator in agriculture such as watermelon, pumpkins, cabbage and lettuce. Ling et al. [32] developed a four-finger prosthetic hand and embedded hand controller for tomato harvesting robot. The sensing and picking were 95% and 85%, respectively, compared with a previous prototype. Liu et al. [33] developed a multi-sensory end-effector for spherical fruit harvesting robot using a vacuum pressure sensor, distance sensors, proximity sensors and force sensors. A laser cutting system composed of high power fiber coupled laser diode was used for cutting while suction pad device with two finger gripper was used for grasping the spherical fruit.

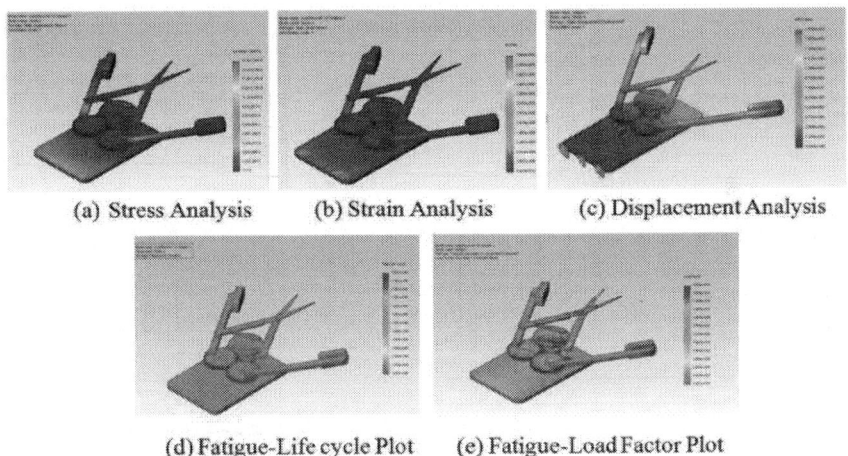

   (a) Stress Analysis   (b) Strain Analysis   (c) Displacement Analysis

    (d) Fatigue-Life cycle Plot   (e) Fatigue-Load Factor Plot

Figure 2. Prototype simulation results of end-effector [28,29].

Bachche et al. [34], Bachche et al. [35] and Bachche and Oka [36] developed a thermal cutting system for sweet pepper harvesting robots based on current and voltage potentials (Figure 3). These systems assist to avoid virus transformation, reduce the fungal vulnerability and increase the shelf life of fruits by adopting a thermal cutting approach. The design consists of two parallel gripper bars mounted on a frame connected by a specially designed notch plate and operated by a servo motor. Based on voltage and current, two different types of thermal cutting system prototypes—electric arc and temperature arc—were developed. In electric arc, a special electric device was developed to obtain high voltage to perform the cutting operation. At higher voltage, electrodes generate thermal arc which help to cut the stem of sweet peppers. In temperature arc, Nichrome wire was mounted between two electrodes and current was provided directly to electrodes which results in generation of high temperature arc between two electrodes that help to perform cutting operation. These prototypes were tested for several variable field conditions in which temperature arc system was found to be effective and took 1.5 s to perform the cutting operation. The post-harvest inspection of harvested fruits confirmed an increase in the shelf life of fruits and prevention of fungal and virus transformation. The fruits harvested by the thermal arc cutting system can be preserved more than 15 days under normal room conditions.

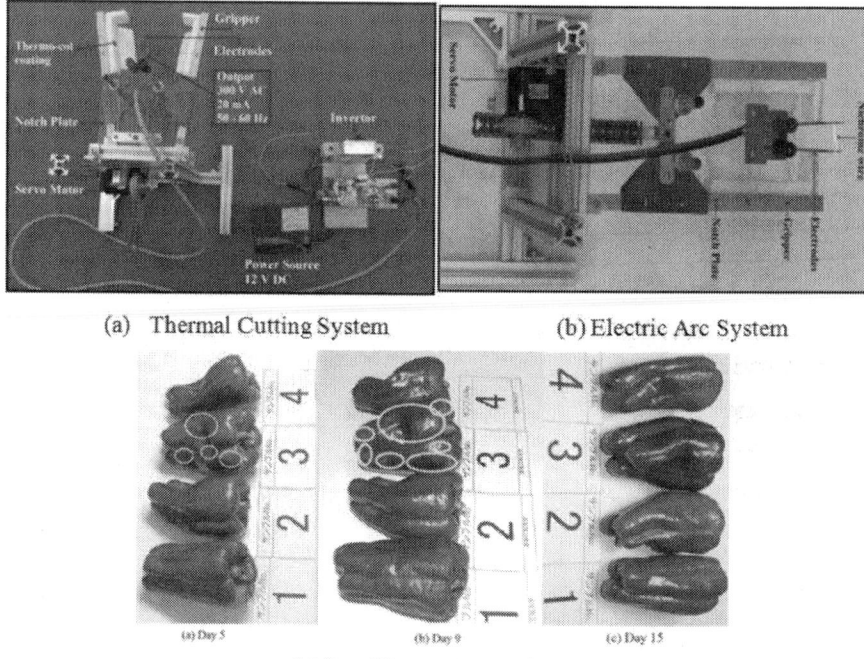

(a) Thermal Cutting System   (b) Electric Arc System

(c) Post Harvest Inspection

Figure 3. Overview of thermal cutting end-effector [34,35,36].

2.2. Recognition System

2.2.1. Color Camera Recognition System

For the last several years, computers have been used extensively for analyzing images and obtaining the data from images. However, due to variability of agricultural objects, it is very difficult to adopt the existing industrial algorithms to agricultural domain. To cope with this variability, the methods and knowledge of algorithms for the agricultural domain need to be studied which could support the variations in field environment conditions and physical flexibility of agricultural objects. There are many processes available in agriculture where decisions are made based on the appearance of the product [37]. The techniques used for these applications are mostly successful under the constrained conditions for which they were designed, but the algorithms are not directly usable in other applications. In principle, computers are flexible because they can be re-programmed, but in practice it is difficult to modify the machine vision algorithms to run for slightly or completely different applications because of the assumptions and rules made to achieve the specific applications [38].

On the other hand, in the agricultural field, configuration of the trees significantly alters the percentage of visible fruits on the tree. For tree row configurations, with a hedge appearance, the visibility of the fruit can reach 75%–80% of the actual number of fruits which is much better than the 40%–50% of visibility for conventional plantings [39]. One major difficulty in developing machinery to selectively harvest fruits is to determine the location, size and ripeness of individual fruits. These specifications are needed to guide a mechanical arm towards the target object. The computer vision strategies used to recognize a fruit rely on four basic features which characterize the object: intensity, color, shape and texture. Apart from these basic characteristics, many researchers are engaged with developing different approaches to recognize fruits in natural backgrounds. Research work on detection of different fruits and vegetables such as apple [40,41,42,43], cherry fruit [44], cucumber [45,46], orange [47,48], tomato [49,50,51], strawberry [52,53], melon [54,55] and sweet pepper [12,56,57,58] (Figure 4) has been undertaken.

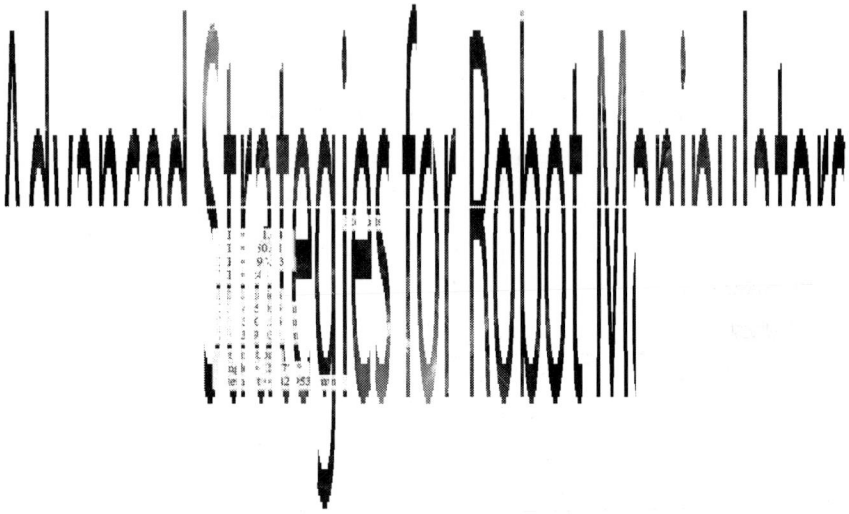

Figure 4. Discrimination and computation of 3D location information of sweet peppers using color CCD cameras based on parallel stereovision principle [12,58]. This positional information of detected green sweet peppers was found to be highly reliable in which the depth accuracy errors and disparity parallax errors were minimal when distance between cameras and fruit was 500–600 mm and distance between two cameras was maintained at 100 mm.

The detailed review on computer vision methods for locating fruits on trees is given by Jimenez, Ceres and Pons [13] which covers the main features of recognition approaches, sensing systems used to capture the images, image processing strategies used to detect the fruits and results obtained from previous studies. Another detailed review was given by McCarthy et al. [59] on applied machine vision of plants with implications for field deployment in automated farming operations in which the research studies conducted previously were grouped into monocular vision with RGB camera, stereo vision and 3D structure, multispectral imaging and range sensing. Each group focuses on the recognition strategies and approaches under a particular group, and potential methods to enhance the machine vision system design for application in the agricultural field were discussed. Kapach et al. [60] provided a comprehensive review of classical and state-of-the-art machine vision solutions employed in harvesting robot systems, with special emphasis on the visual cues, computational approaches and machine vision algorithms used. The studies on image processing approaches for spherical and non-spherical fruits based on visual cues like color, spectral reflectance, thermal response, texture, shape etc. and based on machine vision algorithms like segmentation, clustering, template matching, shape inference, voting, machine learning etc., were discussed in detail according to the applications and algorithms developed considering specific needs for particular applications. Pal, N. and Pal, K. [9] also reviewed studies on image segmentation techniques focusing on fuzzy and non-fuzzy methods for color segmentation, edge detection, surface based segmentation, gray level thresholding and neural network-based approaches. Adequate attention was paid to segmentation of range images, magnetic resonance images and quantitative evaluation of segmentation.

In 1984, Baylou et al. [61] developed a detection system for asparagus using a stereoscopic visual sensor. This system can detect the asparagus and also provide 3D location. Humburg and Reid [62] developed and tested a machine vision system for identification and location of harvestable spears of asparagus in which a videotape of a row of asparagus was used to simulate the vision system. Accumulated errors were found in recognition of spears at confidence interval levels and these intervals were used to determine the size of the cutting mechanism. An effective vision algorithm was presented by Kondo et al. [63] to detect positions of many small fruits and to identify cluster information along with 3D position. The experimental results showed that this visual feedback control based harvesting method was effective, with a success rate of 70%. Edan et al. [64] reported experimental results by replacing the visual system from a melon harvesting robot presented by Edan and Miles [65]. The new vision system consists of black and white image processing system with algorithm that was able to detect fruits and then compute positional information. The intelligent control system consists of a distributed blackboard system with autonomous modules for sensing, planning and control. This robot with a new vision system had a fruit harvesting success rate of more than 85%.

Takahashi et al. [66] reported a method for measurement of 3D location of apples by binocular stereo vision system which had a 90% recognition rate when image was less dense with fruits; and a 65–70% rate when image was denser with red fruits. A machine vision system and laser ranging system was used by Bulanon et al. [67] to determine 3D location of the apples. A laser ranging system was found to be more effective as it determines the distance from the end-effector to the fruit very easily and accurately while the machine vision system was found to be computationally expensive and time consuming. Tarrio et al. [68] reported a method for a recognition system to detect the small fruits in bunches based on 3D stereoscopic vision. The method uses passive and active 3D reconstruction technique, stereoscopic vision and structured lighting. Two CCD cameras with movable laser diode panels were used to illuminate the scene, and color transformation from RGB to HSV was used in image processing.

Lak et al. [69] described apple fruit recognition system under natural luminance using machine vision. An edge detection and combination of color and shape analyses were utilized to segment images of red apples obtained under natural lighting. An edge detection based algorithm was found to be unsuccessful, while color-shape based algorithm could detect apple fruits in 83.33% of images. Li et al. [70] reported development of a real-time fruit recognition system for pineapple harvesting robots in China using color space transformation from RGB to HSI. Ji et al. [71] mentioned a procedure to develop an automatic recognition system based on color and shape features and a new classification algorithm based on support vector machines for apple recognition. The method was found to be efficient with an 89% apple recognition rate and 352 ms average recognition time per fruit. A recognition system for olives using neural networks based on color and shape extracted from captured RGB images was reported by Gatica et al. [72]. Two cases viz., recognition of olives and overlapping of olives, were analyzed during decision making operation based on neural networks.

2.2.2. Multispectral Recognition System

Multispectral imaging is a technique for recognizing and characterizing physical properties of materials using the principle of the varying absorption (or emission) of different wavelengths of light by the objects. This technique has been applied in various areas of science such as medicine, forensics, geology, and meteorology. The wavelengths of light used in multispectral imaging usually lie within the Infrared (IR) and Near Infrared (NIR) ranges. In contrast to hyperspectral imaging, which characterizes materials by measuring the variation in light intensity over continuous ranges of wavelengths, multispectral imaging utilizes a relatively small set of specific wavelengths. The desired wavelengths may be selected by a set of dichroic interference filters of specific wavelength and pass-band. The variation of light intensity versus wavelength is measured by means of appropriate sensor techniques for each point in the scene being imaged. This technique was originally developed for space-based imaging and can allow extraction of additional information the human eye fails to capture with its receptors for red, green and blue. For a different purpose, a different combination of spectral bands can be used represented by red, green and blue channels. Mapping of bands to colors depends on the purpose of the image and personal preferences. For the last few decades, NIR spectroscopy has shown considerable promise for the non-destructive analysis of food products and is ideally suited for on-line measurements in the agro-food industry due to its advantages: minimal or no sample preparation, versatility, speed and low cost analysis [73].

Most well-known applications of NIR spectroscopy in fruits have focused on the quantitative prediction of chemical composition, internal damage and ripening stage in various fruits such as apple, avocado, banana, caraway, coriander, carrot, Chinese bayberry, citrus fruit, dates, dill, fennel, grape, green beans, guava, Japanese pear, kiwifruit, macadamia, mango, melon, mushroom, nectarine, olive, onion, papaya, peach, peach, pepper, plum, tangerine, tomato, etc. The use of multispectral-image-based perception methods has been studied extensively to assess the crop nutrition level based on crop canopy reflectance in multiple spectral bands [74,75]. Optical properties are based on reflectance, transmittance, absorbance, or scatter of light by the product. Often features related with these properties are chosen for various purposes such as external qualities like size, shape, color, gloss, texture, defect; internal qualities like flavor, texture, nutrition, defect, pH level, total soluble solids content, dry matter, sugar content, nitrogen level, starch, moisture content, essential oil content, different acids, chlorophyll, etc. The detailed review on the determination of various non-destructive measurement of fruits and vegetables quality by means of NIR multi-and hyperspectral imaging techniques; different spectrophotometer designs and measurement principles; other spatial techniques for quality measurements is given by Osborne and Hindle [73], Abbott [76] and Nicolai, et al. [77]. The potential of NIR spectroscopy and imaging as rapid, non-destructive and multi-parametric technique is supported by its applications for fruit and vegetable quality measurement including not only food industry, post-harvest applications but also horticultural crops.

In last few decades, in NIR techniques, most of the work has been done either during quality determination, quality inspection and measurement applications or for post-harvest applications as mentioned in the above section. Though the NIR technique has shown significant results for non-destructive quality measurements of fruits and vegetables delivering satisfactory application development for quality measurement, inspection and post-harvest operations, still there has not been enough research done on NIR multispectral imagining for discriminating the fruits in the natural background. Czarnowski and Cebula [78] investigated the spectral properties of red, green, yellow and cream colored sweet peppers. The relationship between spectral properties of fruits and leaves was studied and the results showed that up to 700 nm of the fruit and leaves had almost same reflectance and absorbance of light while from 700 nm to 1100 nm, there was a significant difference between reflectance and absorbance. Van Henten et al. [79] studied the possibilities of adoption of spectral properties to detect cucumbers with natural background. Two monochrome cameras were used with 850 nm and 970 nm band-filters, simultaneously. The spectral band filters showed better detection results as a significant difference was found between the spectral properties of cucumber and leaves. Hemming [80] used the same vision system [79] to study the possibility of detection for sweet peppers. The study reported that spectral vision system can be used successfully for detection of sweet peppers.

Safren et al. [81] used hyperspectral imaging for detection of green apples in green natural background. A multistage algorithm was developed that uses several techniques, such as principle components analysis (PCA) and extraction and classification of homogenous objects (ECHO) for analyzing hyperspectral data, as well as machine vision techniques such as morphological operations, watershed, and blob analysis. The recognition rate was reported as 88.1% and error estimated as 14.1% due to overlapping of fruits. Rath and Kawollek [82] performed experiments on robotic harvesting of Gerbera Jamesonii based on detection and 3D modeling of cut flower pedicels. Two high-resolution CCD cameras with near-infrared filters were used for image capturing. From the data of both images and eight plant positions, three-dimensional models of the pedicels were created by triangulation. The evaluated 3D model was used to calculate spatial coordinates for the applied robot control. Based on the results, a pneumatic harvest grabber was developed, which harvested the pedicels by cutting them off. The pedicels harvest rate was recorded as 80%. Using binocular stereovision and spectral imaging, the possibilities of cucumber detection were verified by Yuan et al. [83]. A stereo vision system was used to capture monochrome near-infrared images. A fruit detection algorithm was used in image processing and 3D location was computed using a triangulation model. The fruit recognition rate was found as 86% while the distance errors of grasping position were reported as less than 8.6 mm.

Bulanon et al. [84] used a CCD camera with six band pass filters to examine possibilities of multispectral imaging for citrus fruits detection. Accordingly, 600 nm, 650 nm and 700 nm band pass filters were found to be suitable to discriminate citrus fruit. The latest work in multispectral imaging is reported by Bachche [85] in which a multispectral recognition system was developed using 780, 800, 900 and 960 nm infrared optical filters to investigate the relationship between spectral properties of green sweet peppers; the recognition rate, fruit visibility percentage and maturity determination of recognized sweet peppers under variable field conditions. A 960 nm wavelength was found to be feasible and effective to discriminate the sweet peppers as, at this wavelength, there was significant difference observed between reflectance and absorbance of light. Also, the reflectance of chlorophyll content at this wavelength was used as a correlation factor with saturated red wavelength adsorption region. The histogram ratios based on transformed chlorophyll adsorption ratio index and principle component analysis was used during multispectral image processing. The results confirmed higher recognition rates, maximum fruit visibility percentage and 76% true maturity determination decisions of recognized sweet peppers (Figure 5). Bac et al. [86] reported a quantitative type study to classify different plant parts under variable lighting using multispectral system. This system was specially developed for detection of sweet peppers in the greenhouse using multispectral cameras. The image processing algorithm provides pixel based classification of plant parts under varying light.

3. FRUIT HARVESTING ROBOTS

Among the various farm management operations, harvesting is an important operation which needs not only labor power but also high energy input with high resources. Most of the other farm management operations can be carried out by highly precise and accurate commercialized mechanization techniques but the harvesting operation still has not gained the similar commercialization status to encourage researchers to study and develop agricultural robot applications for harvesting purposes. The study of agricultural robot applications for plant production presumably started with a mechanical citrus harvesting system in 1968 [87].

In 1984, Japanese researchers Kawamura et al. [88] developed the first fruit harvesting robot in Kyoto University, Japan to harvest tomatoes. This robot consists of a 5 DOF manipulator; end-effector; stereovision and battery car as a travelling device. In the following year, Spain and France together started the MAGALI project [40] which focuses on developing a fruit harvesting robot to harvest apples. Under the MAGALI project, several versions of a harvesting robot were built and tested successfully. Prussia [89] reported the ergonomic aspects of manual harvesting which focuses on several ergonomic principles that relate to manual harvesting and emphasizes the merit and demerits of manual harvesting by analyzing visual acuity, color sensitivity, strength, capacity, and productivity.

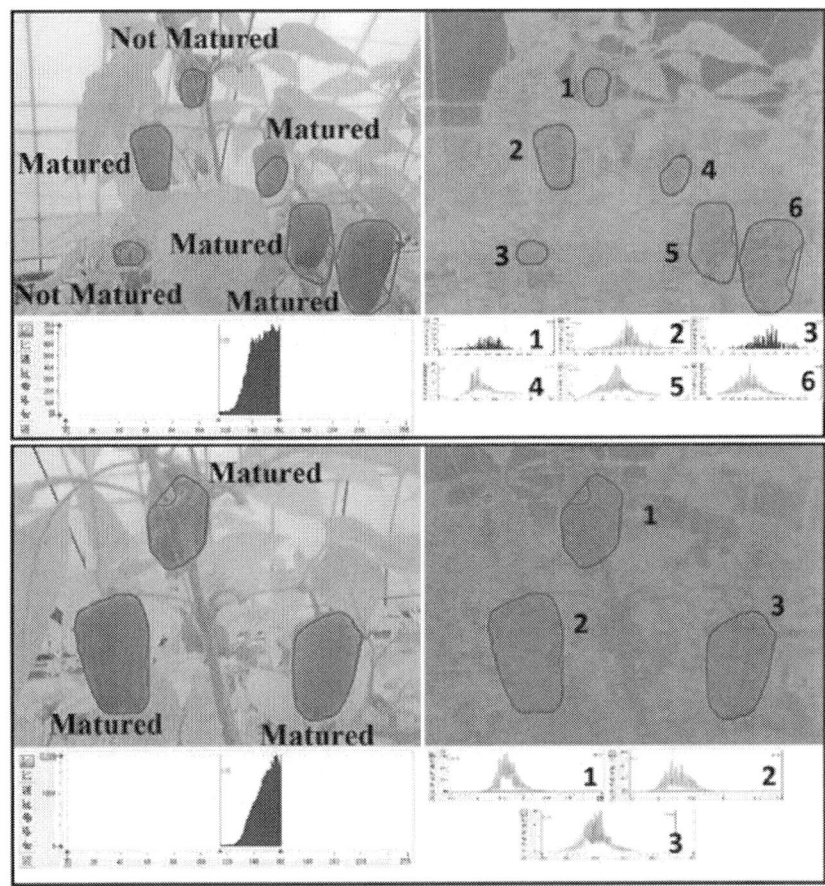

Figure 5. Results of multispectral recognition and maturity determination system [85]. (First image shows pre-known status of detected fruits while second image shows result image with histograms for maturity determination: green color—matured; other colors—not matured).

The first review on robotics and intelligent machines was given by Sistler [90] which highlights the important achievements in mechanization and deliberates on the possibilities of practical agricultural robots. Further, various problems that researchers need to solve, different barriers in development of agricultural robots and how we can increase agricultural productivity by examining the capabilities of intelligent machines and robots are described in detail. Grand D'Esnon et al. [91] reported a second version of MAGALI robot which consists of a spherical manipulator with a vacuum gripper and camera. Four ultrasonic telemeters were used for navigation during harvesting and hydraulic actuators were used to move the manipulator. In 1989, extensive work was performed in fruit harvesting robots: Amaha et al. [92] developed a cucumber harvesting robot at the University of Tokyo, Japan; Whitney and Harrell [93] developed a robotic arm which had a picking rate of one fruit every 5 seconds from outer canopy; an Italian company built a fruit harvesting prototype for citrus which had a 65% harvesting success rate [94]; Sevilla et al. [95] conducted feasibility study of harvesting robot for grapes in France.

Tillett [96] reported a mechatronic mushroom harvester developed in United Kingdom which consists of a black and white vision system, computer controlled Cartesian robot and an end-effector. This robot had an 84% fruit detection rate and 57% harvesting success rate. Sandini et al. [97] developed an autonomous robot for several operations in greenhouses. This robot uses two PAL color cameras with bit-slice microprocessor card for fast image processing. This research was mainly focused on locating the tomatoes in greenhouses using a visual feedback system and navigating the robot through greenhouses to perform several simple operations. Harrell, Adsit, Pool and Hoffman [47] developed a mobile grove-lab to study the use of robotic harvesting of citrus under actual production conditions. This robotic arm was an operational multiple arm harvesting system equipped with several sensors and actuators to perform real-time tasks and controlled through a computer program.

Kubota Co., Sakai, Japan [98] developed a 4 DOF manipulator to harvest oranges. A vacuum pad with an optical proximity sensor in gripper and a stroboscope light with color camera protected by fork shaped cover was used in this manipulator. Pool and Harrell [99] reported developments in a previously built prototype for citrus harvesting. This new design consists of a cutting device with rotating lips that detaches the fruit already enclosed in tube which was attached to a tubular arm for picking soft fruits. This prototype showed a 69% harvesting success rate while 37% of fruits were physically damaged. Kassay [41] reported first Hungarian AUFO robot for harvesting apples. This AUFO robot consists of two color cameras for automatic fruit detection and 3D position of fruit using stereovision system; six arms with movement in vertical plane; four padded fingers with compensation springs to grip the fruit at an accurate grasping pressure; and robotic mobile platform to navigate the robot around the trees. In Israel, Benady and Miles [100] manufactured a melon harvesting robot with Cartesian manipulator mounted on frame moved by a tractor. The robot near-vision system was equipped with laser line projector which illuminates the scene and based on curvature shape where light contacts with melon, using profile transformation and triangulation method, melons were detected and positional information was obtained. To harvest melons at a successful rate, Edan and Miles [65] presented animated, visual and numerical simulations to optimize picking time, actuator speed and planting distance between melons.

Sarig [101] described a detailed review of technical developments and different aspects of harvesting problems in different countries from 1982 to 1992. Tillett [102] reviewed robot manipulators used to handle biological materials within the horticultural industry. The potential applications were discussed and the scope for future developments based on economic as well as technical considerations was deliberated. The developments in locating and picking performance of mushroom harvesting robot were discussed by Reed and Tillett [103]. A black and white vision system with mushroom locating image analysis algorithm, a computer controlled Cartesian robot and a specialized mushroom picking end-effector was used to test the robot's performance. In total, 84% of mushrooms were located by image analysis algorithm while 57% of mushrooms were harvested successfully. Under an Italian Project, Grattoni et al. [104] reported a mobile robot equipped with a suitable manipulator and driven by a stereo-vision module to harvest asparagus.

The development of grape harvesting robot in Japan was reported by Monta et al. [105]. This robot consists of a manipulator, a visual sensor, a travelling device and end-effectors. The visual sensors locate the peduncle of grape cluster and then end-effector grasp and cut peduncle. Further, a bagging system was also developed to cover the cluster and then grasp and cut the peduncle and then transfer to a container. A Spanish project was conducted to develop a human assistive robotic system for greenhouse operations known as Agribot [106]. The operator using a joystick moves a laser pointer until the laser spot is in the middle of the fruit which helps to obtain 3D position of fruit using a computer. Then, a manipulator with pneumatic gripping system and optical proximity sensor which had the ability to detach fruits was moved towards the target fruit. On the other hand, Edan [107] reviewed several developments in autonomous agricultural robots including guidance systems, greenhouse autonomous systems and fruit harvesting robots. A general concept for a field crops' robotic machine to selectively harvest easily bruised fruit and vegetables was presented in this review and future trends that must be pursued in order to make robots a viable option for agricultural operations were discussed.

Several types of fruit harvesting robots were reported by Kondo, Monta and Fujiura [49] such as tomato, petty-tomato and cucumber, and robotic systems such as manipulators, end-effectors, visual sensors and travelling devices were discussed. Arima, Kondo and Monta [45] developed a robotic system for cucumber harvesting which consists of a 6 DOF articulated manipulator, a monochrome TV camera with optical filter having a wavelength of 850 nm and a peduncle detector type end-effector. The visual sensor discriminates the green fruits from green background based on morphological characteristics and an end-effector harvests the fruit. Meanwhile, a Spanish project developed an autonomous mobile robot for greenhouse operations known as AURORA [108]. This was a multi-tasking, remotely supervised AND controlled robot developed for greenhouse operations governed by a control architecture that supports both autonomous navigation and shared human control. This robot has been successfully tested in different greenhouses for autonomous navigation and spraying tasks.

Arndt et al. [109] discussed the trends and developments of an automated selective asparagus harvesting robot that has been operational since 1989 at the Centre for Advanced Manufacturing and Industrial Automation (CAMIA), University of Wollongong, Australia. CAMIA prototype showed effective results in harvesting and had a 94% success rate of harvesting. The specific designs and experimental results of AgriBot were presented by Ceres et al. [110]. The robot was tested in the laboratory with artificial trees to check the performance. The overall performance by AgriBot was found to be effective and satisfactory; the gripping-cutting cycle was 2 s; torque requirement was reduced by 3%. Kondo and Monta [111] developed two types of strawberry harvesting robots: one for hydroponic system and the other for soil systems. The robot consists of a 3 DOF manipulator, aspirating-type pneumatic end-effectors and visual sensors. The pneumatic end-effector grasps the fruit with pressure and then twists the peduncle which results in detachment of fruit.

Reed et al. [112] reported a robot that was capable of automatic mushroom detection, sizing, selection, picking, trimming, conveying and transfer of mushroom. A suction cup end-effector was designed for picking the mushrooms while flexible fingers, high-speed knives and padded pneumatic gripper was designed to handle delicate operations such as conveying, trimming and transferring. A monochrome video camera equipped with 32 W circular fluorescent lighting system and image processing algorithms were used for recognition of mushrooms. This robot showed effective performance under selected operations with 80% of average picking rate. Brown [113] performed labor productivity and harvest cost analysis for citrus based on results obtained from eight mechanical harvesting methods. Van Henten, Hemming, Van Tuijl, Kornet, Meuleman, Bontsema and Van Os [79] from Netherlands developed an autonomous robot for harvesting cucumbers. This robot consists of 7 DOF manipulator, autonomous vehicle, end-effector, two computer vision systems for detection and 3D imaging of the fruit and a control scheme that generates collision-free motions for the manipulator during harvesting. Thermal cutting device was used in this robot to prevent spreading of viruses. The robot had 95% detection rate, 80% harvesting success rate and required 45 s to pick one fruit. At the same time, in Japan, Hayashi et al. [114] developed eggplant harvesting robotic system which includes a machine vision algorithm combining a color segment operation, a visual feedback fuzzy control system for manipulator and peduncle cutting mechanism with gripper. This robot showed 62.5% successful harvesting rate and took 64.1 s to cut one fruit. Cho et al. [115] from South Korea developed a 3 DOF lettuce harvesting robot using machine vision and fuzzy logic control. This robot consists of a manipulator with end-effector, a lettuce-feeding conveyor, an air blower, a machine vision system, six photoelectric sensors and fuzzy log controllers. The robot showed effective performance with 94.12% harvesting success rate and took 5 s to cut the lettuce.

Van Henten et al. [116] further presented field tests of a cucumber harvesting robot reported in 2002 with variable field conditions in which the robot took a cycle time of 124 s per harvested cucumber and the average harvesting success rate was found to be 74.4%. Arima et al. [117] developed a Cartesian coordinate type 4 DOF manipulator with suction cup and visual sensor to harvest strawberries on a table-top structure. Hannan and Burks [48] described the main challenges related to fruit detection and robotic harvesting of citrus. The mechanical designs required in fruit harvesting robots and uses of visual systems for future developments were highlighted. Burks et al. [118] provided primarily a literature survey and synthesis which tries to identify the key issues that robotic system developers and horticultural scientists should consider for optimizing plant-machine system performance. Sanders [8] reviewed details of various aspects of orange harvesting systems such as fruit selection, methods of fruit removal, mechanical harvesting systems, manual picking and orchard arrangement. This review was illustrated by the large amount of research data and analysis based on previous work and reviews on orange harvesting.

Kitamura and Oka [56] developed a sweet pepper harvesting robot which was a sliding rail type timing gear based prototype. The robot consists of mobile base controlled by S-box; a vision system with artificial lighting; HSI binarization algorithm for image processing and picking system consisting of parallel metal fingers and pruner. During harvesting experiments, 29 s were taken by prototype to perform harvesting cycle per fruit and judgment errors were reported due to overlapping of fruits. Kondo et al. [119] updated the results for a strawberry harvesting robot developed in 2004. The updates focus on development of new end-effector techniques for grasping and cutting operations. Muscato et al. [120] proposed three end-effector designs for an orange picking robot. First design uses a mirror placed beneath fruit to assist the camera in moving the end-effector to be centered around the fruit and a de-pressurizable tube used to cut and hold the fruit. This end effector was never constructed because of concerns over cost and robustness. The second design prototype consists of three pneumatically activated flexible fingers which grasp the fruit by pneumatic pressure. A force sensor based circular micro-saw cutter was attached to wrist of manipulator to cut the stem of grasped fruit. This prototype showed a high success rate but the cost of implementation was too high to be used in a commercial model. The third design prototype involved two jaws which grasp the fruit and guide the stem to V-shape cutter located on upper jaw; when jaws closed, the fruit detaches from the plant.

Foglia and Reina [121] developed agricultural robot for radicchio harvesting that consists of a double four-bar linkage manipulator, a special gripper and computer vision to localize the plants in the field based on intelligent color filtering and morphological operations. The performance of the computer vision system was analyzed in terms of accuracy, robustness to noises, and variations in lighting. Belforte et al. [122] developed a multi-tasking robotic system which was able to perform several tasks in greenhouse. This work also illustrates the precise operations on precision spraying and precision fertilization and discusses the important features and requirements for robots in a greenhouse.

Further, Ota, et al. [123] performed experiments with a cucumber leaf picking device which consists of a picking rotor composed of knives and brushes, a motor and a vacuum cleaner. The smooth cut surface of the leaf stalk and the smooth cut surface with small skin was 90% when rotor was configured with two knives and two brushes having insertion speed of 50 mms^{-1}. The average execution time per leaf was 1.1–1.3 s which was much higher than the de-leafing robot developed by Van Henten et al. [124]. Kondo et al. [125] developed an end-effector and manipulator control system for a tomato cluster harvesting robot. The robot moves towards the main stem and grabs it followed by a gripping and cutting action by the end-effector. A pushing device was used at the end-effector to hold the harvested tomato cluster and transfer it to the container without vibrating it. A quick motion control with modified input shaping method considering natural frequency of fruit cluster was used to dampen the vibrations.

Tanagaki et al. [126] developed a cherry harvesting robot which consists of a 4 DOF manipulator, 3D vision system, an end-effector, a computer control program and a travelling device. 3D vision sensor was equipped with red and infrared laser diodes which scan object simultaneously. By processing the images from the 3-D vision sensor, the locations of the fruits and obstacles were recognized, and the trajectory of the end effector was determined. Baeten et al. [127] designed an autonomous apple harvesting robot using 6 DOF industrial manipulator, silicon funnel gripper and camera mounted inside the gripper. A three stage approach was used during harvesting operation. The apple harvesting success rate was 80% with 8-10 s time to pick one apple.

Irie et al. [128] developed asparagus harvesting robot coordinated with 3D vision sensor. A telescopic robot hand which includes set of DC motors and 3D vision sensor used to grip and cut the plants in greenhouse. The cycle time for harvesting operation was recorded as 13.9–23.9 s. Van Henten et al. [129] illustrated optimal manipulator design for a cucumber harvesting robot using task specific manipulation technique and parameter optimization simulations. Several types of robots with different link parameters were analyzed for performance index and a 4 link PPRR type manipulator was found effective. Scarfe et al. [130] reported development of an autonomous kiwifruit picking robot based on intelligent vision system. The robot receives instruction by radio link and operates autonomously as it navigates through the orchard, picking fruit, unloading full bins of fruit, fetching empty bins and protecting the picked fruit from rain. The robot has 4 picking arms, each of which will pick one fruit per second. Chatzimichali et al. [131] used an integrated robotic system that able to move in the field, identify white asparagus stems, grasp them and then cut accordingly without any physical damage.

Aljanobi et al. [132] used ready-made industrial manipulator having 6 DOF to harvest the dates. Hayashi et al. [133] evaluated strawberry harvesting robot which consists of cylindrical manipulator, end-effector with suction device, machine vision unit, storage unit and travelling unit. The fruit detection rate was recorded as 60% while successful harvesting rate was recorded as 41.3% with suction device and 34.9% without suction device and picking cycle time was noted as 11.5 s. De-An et al. [134] presented design and control of apple harvesting robot using PRRRP structure manipulator, spoon-shaped end-effector with pneumatic actuated gripper and image based vision servo control system. A fruit recognition algorithm based on support vector machine was applied to detect apple. The harvesting success rate was recorded as 77% with average picking time as 15 s per apple. Kohan et al. [135] developed Rosa Damascena harvesting robot using stereoscopic machine vision. A 4 DOF manipulator with stereovision technique was used and relation between camera-flower distance and camera-camera distance was evaluated. It was found that an increase in the distance between the cameras reduces the stereoscopic error, while the increase in the distance between the cameras and the flowers increases the error. The overall harvesting success rate was noted as 82.22%. Li et al. [136] reviewed fruit harvesting methods for fruit harvesting robots. A large research data is categorized according to methods of harvesting, machine vision systems and image date analysis.

Feng et al. [137] mentioned a new strawberry harvesting robot for elevated-trough culture using sonar camera sensor and autonomous navigation system. A 6 DOF industrial manipulator was used with a gripping and cutting tool. The successful harvesting rate was recorded as 86% with a harvesting time of 31.3 s on average and an average error for fruit location was found to be less than 4.6 mm. Wang et al. [138] reported design and co-simulation for tomato harvesting robot using a 4 DOF manipulator and machine vision servo system. The viability and validity of the tomato harvesting robot was preliminarily confirmed by co-simulation for the electromechanical system. Yang et al. [139] described design and experiments of an intelligent monorail cucumber harvesting system. An intelligent harvester runs on monorail and carry harvest box, monorail assembly bracket system and control system. An infrared sensor and intelligent camera used in image processing and matured cucumbers can be distinguished by using gray transformation algorithm, trimming image edge algorithm, and locally maximal variance between-class threshold algorithm. The harvest success rate was reported as 97.28% when operating speed was 0.6 m/s. Hemming et. al. [140] reported a robot for harvesting sweet-pepper in greenhouses in which a 9 DOF redundant manipulator, two 5 megapixel 2/3" CCD RGB color cameras, a 3D time of flight camera and artificial lighting rig were used. Two different types of end-effector for detaching sweet peppers were used. The first end-effector had a combined grip and cut mechanism while the second end-effector first stabilized fruit using suction cups after which two lips enclose the fruit and cut the fruit peduncle. The fruit detection and localization was performed at two different levels using color and 3D time of flight cameras. The modular robotic system was tested under simplified laboratory conditions in which the detection rate was 97%, localization rate was 86% and harvesting efficiency was 79%. In a commercial greenhouse, this system has proven the ability to harvest sweet peppers autonomously.

Bac et al. [141] analyzed the state-of-the-art and provided insight on future perspectives for harvesting robots in high-value crops. This review article emphasizes harvesting robots regarding crops harvested in a production environment, performance indicators, design process techniques used, hardware design decisions and algorithm characteristics. The current challenges and limitations in developing commercial fruit harvesting robots were discussed and directions for future harvesting robots were deliberated. The article reports on average success rates for localization, detachment, harvest, fruit damage, peduncle damage and cycle time as 85%, 75%, 66%, 5%, 45% and 33 s, respectively, for the 50 selected different types of fruit harvesting robots.

The timeline of development of fruit harvesting robots around the world can be seen in Figure 6 which shows the horticultural product and the country where it was developed. In the 90s, considerable work has been carried out in Japan and USA on fruit harvesting robots, while in the 21st century, many other countries also started to contribute to this field.

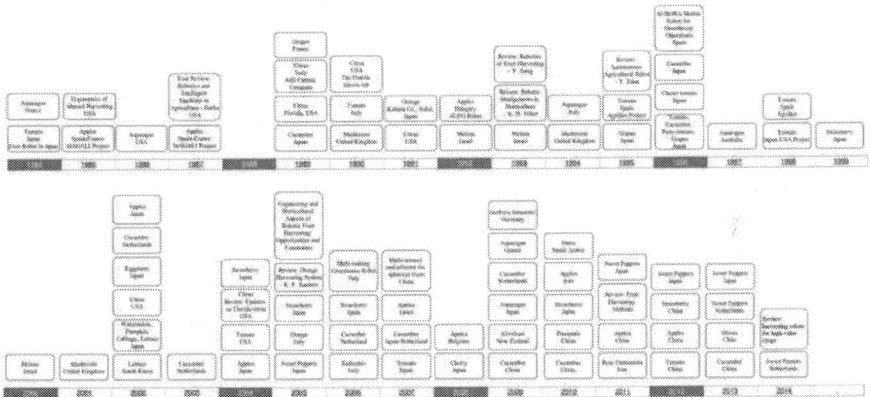

Figure 6. Timeline for research work in fruit harvesting robots around the world.

4. DISCUSSION

Over the centuries, agriculture has transformed into the modern bio-industry it is today, something inconceivable to humans when agriculture started in traditional hunter-gatherer societies. The major changes in agriculture have occurred through domestication of mechanization, use of modern high-tech sophisticated farm management techniques, adoption of cutting edge technologies and pushing engineering to its utmost limits for precision farming and protected cultivation. The changes and developments in agriculture have been seen around the globe in terms of incredible revolutions, and various innovations and developments of machines and robots over time. The application of robots in horticultural product harvesting has also shown very significant and promising results; especially when the agriculture labor population is decreasing with an increase in labor wages and increasing harvesting energy consumption.

So far, considerable and remarkable research work has been carried out around the world, and still many researchers are engaged in finding solutions for several specific task-oriented problems and developing cutting edge technologies which will help the growth of modern agriculture. On the other hand, no harvesting robot has reached the stage of commercialization due to several problems and difficulties such as their low operation speed, low fruit recognition rate in variable field conditions, low rate in obtaining spatial information on fruit, low success rate in fruit harvesting, complexities in the robot manipulator movements, complicated manipulator control methods, difficulties in mobile navigation of robots, problems in simultaneous see-grasp-cut operations, accuracy-repeatability-cycle time gaps and the high costs that are involved in research and manufacturing. Besides these problems, there are several other problems concerning the physical and morphological properties of fruits, horticultural constraints and variable field conditions.

To overcome these problems and to obtain the automatic fruit harvesting system in the agriculture sector, ideally, three main problems need to be solved [13]: (1) the guidance of the robot through the crops, (2) the location and characterization of the fruit on the trees, and (3) the grasping and detachment of each piece. The first problem is not critical and can be solved using one operator to guide the robot through the crops or adopting line tracing moving base system. The other two problems have received remarkable attention during the last 30 years, although no commercial harvesting robot is available. To solve problems of location and characterization of fruit on trees, an efficient recognition system is required that can locate the fruits on trees with their positional information, i.e., the location and orientation of the fruit. Further, the recognition system should be able to locate the occluded or fruits partially covered by leaves in the variable field environments. Also, to solve the problem of grasping and detachment of fruits, an effective gripping and cutting system is required that can harvest the fruits under various conditions without causing any physical damage to the fruit. Moreover, the picking system should be able to handle the soft, delicate fruits during harvesting time with respect to their various shapes and sizes without causing any damage to trees, and also be able to perform the harvesting operation at higher speeds and very precisely.

As the research methods and approaches presented to solve discrimination problems of green fruit provide significant insights on recognition and computation of positional information of green fruits in natural backgrounds, adopting this type of research for other agricultural fruits has three distinct approaches: first, determining the optimal color-space model for each individual fruit; second, applying the same color-space model for all agricultural products as a universal solution and third, combining the color-space models that have a significant effect on color attributes and features of fruits to be detected. In the first case, the fruit recognition rate will be largely increased while the method will be time consuming as it needs considerable time to collect images of each individual fruit, process and analyze the data and draw conclusions. Applying this type of research for major agricultural fruits will widen the scope for fruit harvesting robots. In the second case, the time can be saved but the recognition rate will be decreased and accuracy of detection will be low as every fruit demonstrates considerable distinctions in physical and chemical properties. In the third case, the search for an optimal combination of color spaces will be a tough challenge for researchers and relies on sophisticated research which will not only increase the detection accuracy but also could be used as a universal solution to detect the fruits in natural backgrounds.

In agriculture, fruits and vegetables demonstrate great diversity in their properties and, due to that, researchers need to design and develop different systems for each product. Application of one type of robot designed for a specific product was not feasible for another product. This problem is common to many developments in science and technology, and there are hopes to create some multi-tasking and multi-sensory devices. So, instead of developing separate robotic systems for each product, researchers should try to find out a universal 'one size fits all' solution. This type of work of course needs time, money, lots of research and much more, but this will be an extreme engineering miracle in agriculture. For example, in recognition systems, based on feature attributes, using a single multispectral or hyperspectral vision system with intelligent image processing algorithm, detection and obtaining spatial information for several types of fruits or vegetables is possible. Using one color camera and one multispectral camera [142], the fruits could be recognized successfully and also the maturity status of respective fruits could be determined. This will ensure that only matured fruits could be harvested using robots so as to reduce the operational waste and increase yielding capacity. For end-effectors, developing multi-tasking and multi-sensory end-effector, performing several operations in a sequential loop is possible. So far, researchers are working to develop gripping systems to grasp fruit or fruit clusters only, and this type of gripping requires special care to avoid physical damage to fruits based on various sizes and shapes. On the other hand, this situation could be changed by developing multi-sensory grippers to grasp peduncle of the fruit or fruit cluster instead of grasping fruits directly. This type of devices fits with almost all types of fruits or fruit clusters and also helps to avoid physical damage to fruits. A combination of these types of robotic systems will help to create real-time intelligent robotic systems for fruit harvesting. The real-time cost effective and fully automatic robotic fruit harvester might have come a long way from the final commercial prototype; however, there is still some room in future research to develop these automatic fruit harvesters or harvesting systems

REFERENCES

1. FAO. Core data statistics on agriculture labor population in the world. 2012. Available online: http://www.faostat.fao.org (accessed on 10 August 2012).

2. Kondo, N.; Ting, K.C. Robotics for plant production. Artif. Intell. Rev. **1998**, 12, 227–243.

3. Kondo, N.; Monta, M. Fruit harvesting robotics. J. Robot. Mechatron. **1999**, 11, 321–325.

4. Hashimoto, Y. Agro-robotics. J. Robot. Mechatron. **1999**, 11, 171–172.

5. Ferguson, L.; Rosa, U.A.; Castro-Garcia, S.; Lee, S.M.; Guinard, J.X.; Burns, J.; Krueger, W.H.; O'Connell, N.V.; Glozer, K. Mechanical harvesting of california table and oil olives. Adv. Hortic. Sci. **2010**, 24, 53–63.

6. Coppock, G.E. Harvesting early and midseason citrus fruit with tree shaker. Fla. Agric. Exp. Station. J. Ser. **1967**, 2824, 98–104.

7. Holt, J.S. Implications of reduced availability of seasonal agricultural workers on the labor intensive sector of us agriculture. In Proceedings of ASAE Annual International Meeting, Toronto, Canada, 18–22 July 1999.

8. Sanders, K.F. Orange harvesting system review. Biosyst. Eng. **2005**, 90, 115–125.

9. Pal, N.; Pal, K. A review on image segmentation techniques. Pattern Recogn. **1993**, 26, 1277–1294.

10. Jimenez, R.; Jain, A.K.; Ceres, R.; Pons, J.L. Automatic fruit recognition: A survey and new results using range/attenuation images. Pattern Recogn. **1999**, 32, 1719–1736.

11. Radke, R.J.; Andra, S.; Al-Kofahi, O.; Roysam, B. Image change detection algorithms: A systematic survey. IEEE Trans. Image Process. **2005**, 14, 294–307.

12. Bachche, S.; Oka, K. Distinction of green sweet pepper by using various color space models and computation of 3 dimensional coordinates location of recognized green sweet peppers based on parallel stereovision system. J. Syst. Des. Dyn. **2013**, 7, 178–196.

13. Jimenez, A.R.; Ceres, R.; Pons, J.L. A survey of computer vision methods for locating fruit on trees. Trans. ASAE **2000**, 43, 1911–1920.

14. Blanes, C.; Mellado, M.; Ortiz, C.; Valera, A. Technologies for robot grippers in pick and place operations for fresh fruits and vegetables. Span. J. Agric. Res. **2011**, 9, 1130–1141.

15. Montana, D.J. Contact stability for two-fingered grasps. IEEE Trans. Robot. Autom. **1992**, 8, 421–430.

16. Funahashi, Y.; Yamada, T.; Tate, M.; Suzuki, Y. Grasp stability analysis considering the curvatures at contact points. In Proceedings of the International Conference on Robotics and Automation, Minneapolis, MN, USA, 20–28 April 1996; Volume 4, pp. 3040–3046.

17. Jenmalm, P.; Goodwin, A.W.; Johansson, R.S. Control of grasp stability when humans lift objects with different surface curvatures. J. Neurophysiol. **1998**, 79, 1643–1653.

18. Svinin, M.M.; Kaneko, M.; Tsuji, T. Internal forces and stability in multi-finger grasps. Control Eng. Pract. **1999**, 7, 413–422.

19. Morales, A.; Sanz, P.J.; del Pobil, A.P.; Fagg, A.H. Vision-Based three-finger grasp synthesis constrained by hand geometry. Robot. Auton. Syst. **2006**, 54, 496–512.

20. Birglen, L.; Gosselin, C.M. Grasp-state plane analysis of two-phalanx underactuated fingers. Mech. Mach. Theory **2006**, 41, 807–822.

21. Kragten, G.A.; Herder, J.L.; Schwab, A.L. On the influence of contact geometry on grasp stability. In Proceedings of the ASME 2008 IDETC/CIE, New York, NY, USA, 3–6 August 2008.

22. Aleotti, J.; Caselli, S. Interactive teaching of task-oriented robot grasps. Robot. Auton. Syst. **2010**, 58, 539–550.

23. Noohi, E.; Moradi, H.; Noori, N.; Ahmadabadi, M.N. Manipulation of polygonal objects with two wheeled-tip fingers: Planning in the presence of contact position error. Robot. Auton. Syst. **2011**, 55, 44–55.

24. Daoud, N.; Gazeau, J.P.; Zeqhloul, S.; Arsicault, M. A real-time strategy for dexterous manipulation: Fingertips motion planning, force sensing and grasp stability. Robot. Auton. Syst. **2012**, 60, 377–386.

25. Li, Z.; Li, P.; Yang, H.; Wang, Y. Stability tests of two-finger tomato grasping for harvesting robots. Biosyst. Eng. **2013**, 116, 163–170.

26. Monkman, G.J.; Hesse, S.; Steinmann, R.; Schunk, H. Robot Grippers; Wiley-VCH Verlag GmbH and Co. KGaA Weinheim: Germany, 2007.

27. Edan, Y.; Haghighi, K.; Stroshine, R.; Cardenas-Weber, M. Robot gripper analysis: Finite element modeling and optimization. Appl. Eng. Agric. **1992**, 8, 563–570.

28. Bachche, S.; Oka, K.; Sakamoto, H. Design and modeling of gripper and cutting tool system for sweet pepper harvesting robot hand. In Proceedings of the MAGDA Conference in Pacific Asia, Kaohsiung, Taiwan, 14–16 November 2011.

29. Bachche, S.; Oka, K. Modeling and performance testing of end-effector for sweet pepper harvesting robot. J. Robot. Mechatron. **2013**, 25, 705–717.

30. Monta, M.; Kondo, N.; Ting, K.C. End-effector for tomato harvesting robot. Artif. Intell. Rev. **1998**, 12, 11–25.

31. Sakai, S.; Lida, M.; Umeda, M. Heavy material handling manipulator for agricultural robot. In Proceedings of the IEEE International Conference on Robotics and Automation, Washington, DC, USA, 11–15 May 2002; pp. 1062–1068.

32. Ling, P.P.; Ehsani, R.; Ting, K.C.; Chi, Y.; Ramalingam, N.; Klingman, M.H.; Draper, C. Sensing and end-effector for a robotic tomato harvester. 2004 ASAE Annu. Meet. **2004**.

33. Liu, J.; Li, P.; Li, Z. A multi-sensory end-effector for spherical fruit harvesting robot. In Proceedings of the IEEE International Conference on Automation and Logistics, Jinan, China, 18–21 August 2012; pp. 258–262.

34. Bachche, S.; Oka, K.; Sakamoto, H. Development of thermal cutting system for sweet pepper harvesting robot in greenhouse horticulture. In Proceedings of the JSME Conference on Robotics and Mechatronics, Hamamatsu, Japan, 27–29 May 2012.

35. Bachche, S.; Oka, K.; Sakamoto, H. Development of current based temperature arc thermal cutting system for green pepper harvesting robot. In Proceedings of the Shikoku-section Joint Convention of the Institute of Electrical and related Engineers, Takamatsu, Japan, 29 September 2012.

36. Bachche, S.; Oka, K. Performance testing of thermal cutting system for sweet pepper harvesting robot in greenhouse horticulture. J. Syst. Des. Dyn. **2013**, 7, 36–51.

37. Tillett, R.D. Image analysis for agricultural processes: A review of potential opportunities. J. Agric. Eng. Res. **1991**, 50, 247–258.

38. Jain, A.K.; Dorai, C. Practicing vision: Integration, evaluation and applications. Pattern Recogn. **1997**, 30, 183–196.

39. Juste, F.; Sevilla, F. Citrus: A european project to study the robotic harvesting of oranges. In Proceedings of the 3rd International Symposium on Fruit, Nut and Vegetable Harvesting Mechanization, Denmark-Sweden-Norway, 5–15 August 1991; pp. 331–338.

40. Grand D'Esnon, A. Robot Harvesting of Apples. In Proceedings of Agr-Mation, Chicago, IL, USA, 25–28 February 1985.

41. Kassay, L. Hungarian robotic apple harvester. 1992 ASAE Annu. Meet. **1992**, 92-7042, 1–14.

42. Bulanon, D.M.; Kataoka, T.; Zhang, S.; Ota, Y.; Hiroma, T. Optimal thresholding for the automatic recognition of apple fruits. 2001 ASAE Annu. Meet. **2001**.

43. Tabb, A.; Peterson, D.; Park, J. Segmentation of apple fruit from video via background modeling. 2006 ASAE Pap. **2006**.

44. Tanagaki, K.; Fujiura, T.; Akase, A.; Imagawa, I. Cherry harvesting robot. In Proceedings of the International Workshop on Bio-Robotics, Information Technology and Intelligent Control for Bio-Production Systems, Sapporo, Japan, 9–10 September 2006; pp. 254–260.

45. Arima, S.; Kondo, N.; Monta, M. Development of robotic system for cucumber harvesting. Jpn. Agric. Res. Q. **1996**, 30, 233–238.

46. Van Henten, E.J.; Hemming, J.; Van Tuijl, B.A.J.; Kornet, J.G.; Bontsema, J. Collision-free motion planning for a cucumber picking robot. Biosyst. Eng. **2003**, 86, 135–144.

47. Harrell, R.C.; Adsit, P.D.; Pool, T.A.; Hoffman, R. The florida robotic grove-lab. Trans. ASABE **1990**, 33, 391–399.

48. Hannan, M.W.; Burks, T.F. Current developments in automated citrus harvesting. 2004 ASAE Annu. Inter. Meet. **2004**.

49. Kondo, N.; Monta, M.; Fujiura, T. Fruit harvesting robots in japan. Adv. Space Res. **1996**, 18, 181–184.

50. Kondo, N.; Yamamoto, K.; Yata, K.; Kurita, M. A machine vision for tomato cluster harvesting robot. ASAE Annu. Inter. Meet. **2008**, 5, 3111–3120.

51. Jiang, H.; Peng, Y.; Ying, Y. Measurement of 3-d locations of ripe tomato by binocular stereo vision for tomato harvesting. 2008 ASAE Annu. Inter. Meet. **2008**.

52. Guo, F.; Cao, Q.; Cui, Y.; Masateru, N. Fruit location and stem detection for strawberry harvesting robot. Trans. Chin. Soc. Agric. Eng. **2008**, 24, 89–94.

53. Rajendra, P.; Kondo, N.; Ninomoya, K.; Kamata, J.; Kurita, M.; Shiigi, S.; Hayashi, S.; Yoshida, H. Machine vision algorithm for robots to harvest strawberries in tabletop culture greenhouse. Eng. Agric. Environ. Food **2009**, 2, 24–30.

54. Benady, M.; Edan, Y.; Hetzroni, A.; Miles, G.E. Design of a field crops robotic machine. Pap. ASAE **1991**, 91-7028, 1–7.

55. Dobrusin, Y.; Edan, Y.; Grinshpun, J.; Peiper, U.M.; Hetzroni, A. Real-time image processing for robotic melon harvesting. Pap. ASAE **1992**, 92-3515, 1–16.

56. Kitamura, S.; Oka, K. Recognition and cutting system of sweet pepper for picking robot in greenhouse horticulture. In Proceedings of the IEEE International Conference on Mechatronics and Automation, Niagara Falls, Ontario, Canada, 29 July–1 August 2005; Volume 4, pp. 1807–1812.

57. Kitamura, S.; Oka, K. Improvement of the ability to recognize sweet peppers for picking robot in greenhouse horticulture. In Proceedings of the International Joint Conference on SICE-ICASE, Busan, Korea, 18–21 October 2006; pp. 353–356.

58. Bachche, S.; Oka, K.; Ogawa, N. Distinction of green sweet pepper by using various color space models. In Proceedings of the Annual Conference of the Robotics Society of Japan, Sapporo, Japan, 17–20 September 2012.

59. McCarthy, C.L.; Hancock, N.H.; Raine, S.R. Applied machine vision of plants: A review with implications for field deployment in automated farming operations. Intell. Serv. Robot. **2010**, 3, 209–217.

60. Kapach, K.; Barnea, E.; Mairon, R.; Edan, Y.; Ben-Shahar, O. Computer vision for fruit harvesting robots-state of art and challenges ahead. Int. J. Comput. Vis. Robot. **2012**, 3, 4–34.

61. Baylou, P.; El Hadi Amor, B.; Monsion, M.; CBouvet, C.; Boussau, G. Detection and three-dimensional localization by stereoscopic visual sensor and its application to a robot for picking asparagus. Pattern Recogn. **1984**, 17, 377–384.

62. Humburg, D.S.; Reid, J.F. Field performance for machine vision for selective harvesting of asparagus. Appl. Eng. Agric. **1986**, 2, 2–5.

63. Kondo, N.; Nishitsuji, Y.; Ling, P.P.; Ting, K.C. Visual feedback guided robotic cherry tomato harvesting. Trans. ASAE **1996b**, 39, 2331–2338.

64. Edan, Y.; Rogozin, D.; Flash, T.; Miles, G.E. Robotic melon harvesting. IEEE Trans. Robot. Autom. **2000**, 16, 831–835.

65. Edan, Y.; Miles, G.E. Design of an agricultural robot for harvesting melons. Trans. ASAE **1993**, 36, 593–603.

66. Takahashi, T.; Zhang, S.; Fukuchi, H. Measurement of 3-d locations of fruit by binocular stereo vision for apple harvesting in an orchard. 2002 ASAE Annu. Inter. Meet. **2002.**

67. Bulanon, D.M.; Kataoka, T.; Okamoto, H.; Hata, S. Determining the 3-d Location of the Apple Fruit During Harvest. In Proceedings of the Automation Technology for Off-Road Equipment, Kyoto, Japan, 7 October 2004; pp. 91–97.

68. Tarrio, P.; Bernardos, A.M.; Casar, J.R.; Besada, J.A. A harvesting robot for small fruit in bunches based on 3-d stereoscopic vision. In Proceedings of the World Congress Conference on Computers in Agriculture and Natural Resources, Orlando, FL, USA, 24–26 July 2006; pp. 270–275.

69. Lak, M.B.; Minaei, S.; Amiriparian, J.; Beheshti, B. Apple fruits recognition under natural luminance using machine vision. Adv. J. Food Sci. Technol. **2010**, 2, 325–327.

70. Li, B.; Wang, M.; Wang, N. Development of a real-time fruit recognition system for pineapple harvesting robots. In Proceedings of the Annual Meeting of ASABE, Pittsburgh, PA, USA, 20–23 June 2010.

71. Ji, W.; Zhao, D.; Cheng, F.; Xu, B.; Zhang, Y.; Wang, J. Automatic recognition vision system guided for apple harvesting robot. Comput. Electr. Eng. **2012**, 38, 1186–1195.

72. Gatica, G.; Best, S.; Ceroni, J.; Lefranc, G. Olive fruits recognition using neural networks. Procedia Comput. Sci. **2013**, 17, 412–419.

73. Osborne, B.G.; Hindle, P.H. Practical NIR Spectroscopy with Applications in Food and Beverage Analysis; Longman Scientific: Harlow, Esex, UK, 1993.

74. Kim, Y.S.; Reid, J.F.; Hansen, A.C.; Zhang, Q. On-field crop stress detection system using multispectral imaging sensor. Agric. Biosyst. Eng. **2000**, 1, 88–94.

75. Sui, R.; Wilkerson, J.B.; Hart, W.E.; Wilhelm, L.R.; Howard, D.D. Multi-spectral senseo for detection of nitrogen status in cotton. Appl. Eng. Agric. **2005**, 21, 167–172.

76. Abbott, J. Quality measurement of fruits and vegetables. Postharvest Biol. Technol. **1999**, 15, 207–225.

77. Nicolai, B.M.; Beullens, K.; Bobelyn, E.; Peirs, A.; Saeys, W.; Theron, K.I.; Lammertyn, J. Non-destructive measurement of fruit and vegetable quality by means of nir spectroscopy: A review. Postharvest Biol. Technol. **2007**, 46, 99–118.

78. Czarnowski, M.; Cebula, S. Spectral properties of sweet pepper fruits. Folia Hortic. **1998**, 10, 39–51.

79. Van Henten, E.J.; Hemming, J.; Van Tuijl, B.A.J.; Kornet, J.G.; Meuleman, J.; Bontsema, J.; Van Os, E.A. An autonomous robot for harvesting cucumbers in greenhouses. Auton. Robot. **2002**, 13, 241–258.

80. Hemming, J.; Wageningen University, Wageningen, the Netherlands; Bachche, S.; Tohoku University, Sendai, Japan. Personal communication. 2003.

81. Safren, O.; Alchanatis, V.; Ostrovsky, V.; Levi, O. Detection of green apples in hyperspectral images of apple-tree foliage using machine vision. Trans. ASABE **2007**, 50, 2303–2313.

82. Rath, T.; Kawollek, M. Robotic harvesting of gerbera jamesonii based on detection and three-dimensional modeling of cut flower pedicels. Comput. Electr. Agric. **2009**, 66, 85–92.

83. Yuan, T.; Li, W.; Feng, Q.; Zhang, J. Spectral Imaging for Greenhouse Cucumber Fruit Detection Based on Binocular Stereovision. 2010 ASAE Annu. Inter. Meet. **2010**.

84. Bulanon, D.M.; Burks, T.F.; Alchanatis, V. A multispectral imaging analysis for enhancing citrus fruit detection. Environ. Control Biol. **2010**, 48, 81–91.

85. Bachche, S. Automatic Harvesting for Sweet Peppers in Greenhouse Horticulture. Ph.D. Dissertation, Kochi University of Technology, Kochi, Japan, 2013.

86. Bac, C.W.; Hemming, J.; van Henten, E.J. Robust pixel-based classification of obstacles for robotic harvesting of sweet-pepper. Comput. Electr. Agric. **2013**, 96, 148–162.

87. Schert, C.E.; Brown, G.K. Basic considerations in mechanizing citrus harvest. Trans. ASAE **1968**, 11, 343–346.

88. Kawamura, N.; Namikawa, K.; Fujiura, T.; Ura, M. Study on agricultural robot. J. Jpn. Soc. Agric. Mach. **1984**, 46, 353–358.

89. Prussia, S.E. Ergonomics of manual harvesting. Appl. Ergon. **1985**, 16, 209–215.

90. Sistler, F.E. Robotics and intelligent machines in agriculture. IEEE J. Robot. Autom. **1987**, 3, 3–6.

91. Grand D'Esnon, A.; Rabatel, G.; Pellenc, R. Magali: A self-propelled robot to pick apples. Available online: http://agris.fao.org/agris-search/search.do?recordID=US8853733 (accessed on 15 June 2015).

92. Amaha, K.; Shono, H.; Takakura, T. A harvesting robot of cucumber fruits. Available online: http://agris.fao.org/agris-search/search.do?recordID=US9166423 (accessed on 15 June 2015).

93. Whitney, J.D.; Harrell, R.C. Status of citrus harvesting in florida. J. Agric. Eng. Res. **1989**, 42, 285–299.

94. Blandini, G.; Levi, P. First approaches to robot utilization for automatic citrus harvesting. In Land and Water Use; Dodd, V.A., Grace, P.M., Eds.; A.A. Balkema: Rotterdam, Netherlands, 1989; pp. 1903–1907. ISBN: 9061919800.

95. Sevilla, F.; Sitticihareonchai, F.; Fatou, J.M.; Constans, A.; Brons, A.; Davenel, A. A robot to harvest grape: A feasibility study. SAE Pap. No.: 89-7084. **1989**.

96. Tillett, R.D. Initial development of a mechatronic mushroom harvester. In Proceedings of the International Conference on Mechatronics: Designing Intelligent Machines, Institution of Mechanical Engineers, Cambridge, UK; 1990; pp. 109–114.

97. Sandini, G.; Buemi, F.; Massa, M.; Zucchini, M. Visually guided operations in greenhouse. In Proceedings of the IEEE International Workshop on Intelligent Robots and Systems, Ibaraki, Japan, 3-6 July 1990; pp. 279–285.

98. Hayashi, U.; Ueda, Y. Orange harvesting robot; Kubota Co.: Sakai, Japan, 1991.

99. Pool, T.A.; Harrell, R.C. An end-effector for robotic removal of citrus from the tree. Trans. ASABE **1991**, 34, 373–378.

100. Benady, M.; Miles, G.E. Locating melons for robotic harvesting using structured light. ASAE Pap. No.: 92-7021. **1992**.

101. Sarig, Y. Robotics of fruit harvesting: A state-of-the-art review. J. Agric. Eng. Res. **1993**, 54, 265–280.

102. Tillett, N.D. Robotic manipulators in horticulture: A review. J. Agric. Eng. Res. **1993**, 55, 89–105.

103. Reed, J.N.; Tillett, R.D. Initial experiments in robotic mushroom harvesting. Mechatronics **1994**, 4, 265–279.

104. Grattoni, P.; Cumani, A.; Guiducci, A.; Pettiti, G. Automatic harvesting of asparagus: An application of robot vision to agriculture. In Proceedings of the SPIE 2058, Mobile Robots VIII, Boston, MA, USA, 1 February 1994; pp. 200–210.

105. Monta, M.; Kondo, N.; Shibano, Y. Agricultural robot in grape production system. In Proceedings of the IEEE International Conference

on Robotics and Automation, Nagoya, Japan, 21–27 May 1995; Volume 3, pp. 2504–2509.

106. Buemi, F.; Massa, M.; Sandini, G. Agrobot: A robotic system for greenhouse operations. Robot. Agric. Food Ind. **1995**, 4, 172–184.

107. Edan, Y. Design of an autonomous agricultural robot. Appl. Intell. **1995**, 5, 41–50.

108. Mandow, A.; Gomez de Gabriel, J.M.; Martinez, J.L.; Munoz, V.F.; Ollero, A.; Garci a-Cerezo, A. The autonomous mobile robot aurora for greenhouse operation. IEEE Robot. Autom. Mag. **1996**, 3, 18–28.

109. Arndt, G.; Rudziejewski, R.; Stewart, V.A. On the future of automated selective asparagus harvesting technology. Comput. Electr. Agric. **1997**, 16, 137–145.

110. Ceres, R.; Pons, J.L.; Jimenez, A.R.; Martin, J.M.; Calderon, L. Design and implementation of an aided fruit-harvesting robot (agribot). Ind. Robot Int. J. **1998**, 25, 337–346.

111. Kondo, N.; Monta, M. Strawberry harvesting robots. ASAE Pap. No.: 99-3071. **1999**.

112. Reed, J.N.; Miles, S.J.; Butler, J.; Baldwin, M.; Noble, R. AE— Automation and emerging technologies: Automatic mushroom harvester development. J. Agric. Eng. Res. **2001**, 78, 15–23.

113. Brown, G.K. Mechanical harvesting systems for the florida citrus juice industry. 2002 ASAE Annu. Meet. **2002**.

114. Hayashi, S.; Ganno, K.; Ishii, Y.; Tanaka, I. Robotic harvesting system for eggplants. Jpn. Agric. Res. Q. **2002**, 36, 163–168.

115. Cho, S.I.; Chang, S.J.; Kim, Y.Y. Development of a three degrees-of-freedom robot for harvesting lettuce using machine vision and fuzzy logic control. Biosyst. Eng. **2002**, 82, 143–149.

116. Van Henten, E.J.; Van Tuijl, B.A.J.; Hemming, J.; Kornet, J.G.; Bontsema, J.; Van Os, E.A. Field test of an autonomous cucumber picking robot. Biosyst. Eng. **2003**, 86, 305–313.

117. Arima, S.; Kondo, N.; Monta, M. Strawberry harvesting robot on table-top culture. 2004 ASAE Annu. Meet. **2004.**

118. Burks, T.F.; Villegsa, F.; Hannan, M.; Flood, S.; Sivaraman, B.; Subramanian, V.; Sikes, J. Engineering and horticultural aspects of robotic fruit harvesting: Opportunities and constrains. HortTechnology **2005**, 15, 79–87.

119. Kondo, N.; Ninomoya, K.; Hayashi, S.; Tomohiko, O.; Kubota, K. A new challenge of robot for harvesting strawberry grown on table top culture. 2005 ASAE Annu. Meet. **2005.**

120. Muscato, G.; Prestifilippo, M.; Abbate, N.; Rizzuto, I. A prototype of an orange picking robot: Past history, the new robot and experimental result. Ind. Robot Int. J. **2005**, 32, 128–138.

121. Foglia, M.; Reina, G. Agricultural robot for radicchio harvesting. J. Field Robot. **2006**, 23, 363–377.

122. Belforte, G.; Deboli, R.; Gay, P.; Piccarolo, P.; Ricauda Aimonino, D. Robot design and testing for greenhouse applications. Biosyst. Eng. **2006**, 95, 309–321.

123. Ota, T.; Bontsema, J.; Hayashi, S.; Kubota, K.; Van Henten, E.J.; Van Os, E.A.; Ajiki, K. Development of a cucumber leaf picking device for greenhouse production. Biosyst. Eng. **2007**, 98, 381–391.

124. Van Henten, E.J.; Van Tuijl, B.A.J.; Hoogakker, G.J.; Van Der Weerd, M.J.; Hemming, J.; Kornet, J.G.; Bontsema, J. An autonomous robot for de-leafing cucumber plants grown in a high-wire cultivation system. Biosyst. Eng. **2006**, 94, 317–323.

125. Kondo, N.; Taniwaki, S.; Tanihara, K.; Yata, K.; Monta, M.; Kurita, M.; Tsutumi, M. An end-effector and manipulator control for tomato cluster harvesting robot. 2007 ASAE Annu. Meet. **2007**.

126. Tanagaki, K.; Fujiura, T.; Akase, A.; Imagawa, I. Cherry-harvesting robot. Comput. Electr. Agric. **2008**, 63, 65–72.

127. Baeten, J.; Donne, K.; Boedrij, S.; Beckers, W.; Claesen, E. Autonomous fruit picking machine: A robotic apple harvester. Field Serv. Robot. **2008**, 42, 531–539.

128. Irie, N.; Tagushi, N.; Horie, T.; Ishimatsu, T. Development of asparagus harvester coordinated with 3-d vision sensor. J. Robot. Mechatron. **2009**, 21, 583–589.

129. Van Henten, E.J.; van't Slot, D.A.; Hol, C.W.J.; van Willigenburg, L.G. Optimal manipulator design for a cucumber harvesting robot. Comput. Electr. Agric. **2009**, 65, 247–257.

130. Scarfe, A.J.; Flemmer, R.C.; Bakker, H.H.; Flemmer, C.L. Development of an autonomous kiwifruit picking robot. In Proceedings of the International Conference on Autonomous Robots and Agents, Wellington, New Zealand, 10–12 February, 2009; pp. 380–384.

131. Chatzimichali, A.P.; Georgilas, I.P.; Tourassis, V.D. Design of an advanced prototype robot for white asparagus harvesting. In Proceedings of the IEEE/ASME International Conference on Advanced Intelligent Mechatronics, Singapore, 14–17 July 2009; pp. 887–892.

132. Aljanobi, A.A.; Al-Hamed, S.A.; Al-Suhaibani, S.A. A setup of mobile robotic unit for fruit harvesting. In Proceedings of the IEEE International Workshop on Robotics in Alpe-Adria-Danube Region, Budapest, 24– June 2010; pp. 105–108.

133. Hayashi, S.; Shigematsu, K.; Yamamoto, S.; Kobayashi, K.; Kohno, Y.; Kamata, J.; Kurita, M. Evaluation of a strawberry-harvesting robot in a field test. Biosyst. Eng. **2010**, 105, 160–171.

134.Zhao, A.; Lv, J.; Ji, W.; Zhang, Y.; Chen, Y. Design and control of an apple harvesting robot. Biosyst. Eng. **2011**, 110, 112–122.

135.Kohan, A.; Borghaee, A.M.; Yazdi, M.; Minaei, S.; Sheykhdavudi, M.J. Robotic harvesting of rosa damascena using stereoscopic machine vision. World Appl. Sci. J. **2011**, 12, 231–237.

136.Li, P.; Lee, S.M.; Hsu, H. Review on fruit harvesting method for potential use of automatic fruit harvesting systems. Procedia Eng. **2011**, 23, 351–366.

137.Feng, Q.C.; Wang, X.; Zheng, W.G.; Qui, Q.; Jiang, K. New strawberry harvesting robot for elevated-trough culture. Int. J. Agric. Biol. Eng. **2012**, 5, 1–8.

138.Wang, J.; Zhou, Z.; Du, X. Design for tomato harvesting robots. In Proceedings of Chinese Control Conference, Hefei, China, 25–27 July 2012; pp. 5105–5108.

139.Yang, Z.; Zhang, W.; Zhang, J.; Ji, C.; Li, W. Design and Experiment of Intelligent Monorail Cucumbers Harvester System, Proceedings of ASABE annual meeting, Kansas, MI, USA, 21–24 July 2013.

140.Hemming, J.; Bac, C.W.; Bart, A.J.; van Tuijl, B.A.J.; Barth, R.; Bontsema, J.; Pekkeriet, E. A robot for harvesting sweet-pepper in greenhouses. In Proceedings of the International Conference of Agricultural Engineering, Zurich, Switzerland, 6–10 July 2014.

141.Bac, C.W.; van Henten, E.J.; Hemming, J.; Edan, Y. Harvesting robots for high-value crops: State-of-the-art review and challenges ahead. J. Field Robot. **2014**, 31, 888–911.

142.Fernández, R.; Montes, H.; Salinas, C.; Sarria, J.; Armada, M. Combination of rgb and multispectral imagery for discrimination of cabernet sauvignon grapevine elements. Sensors **2013**, 13, 7838–7859.

Index